Suneel Kumar

Sustainable Rural Tourism in Himalayan Foothills

Environmental, Social and Economic Challenges: A Study of Himachal Pradesh

Springer

Suneel Kumar
Shaheed Bhagat Singh College
University of Delhi
New Delhi, India

ISBN 978-3-031-40097-1 ISBN 978-3-031-40098-8 (eBook)
https://doi.org/10.1007/978-3-031-40098-8

This Springer imprint is published by the registered company Springer Nature Switzerland AG
The registered company address is: Gewerbestrasse 11, 6330 Cham, Switzerland

Paper in this product is recyclable.

Sustainable Rural Tourism in Himalayan Foothills

Preface

Rural tourism exhorts tourists to visit rural areas for dynamic participation in a rural lifestyle. Rural tourism, which started as a sub-niche of eco-tourism, is currently viewed as a significant tourism product due its socio-economic benefits. An increase in employment, development of infrastructure, rising standard of living, as well as exchange of ideas, knowledge, and customs across communities are some of the benefits provided by the tourism industry. Recent studies have highlighted that certain types of rural tourism are not sustainable. They do more damage to nature and surroundings than provide benefits. Therefore, there is an inherent need to introduce specific measures in the development of sustainable rural tourism. Urban destinations have either been exhausted or are on the verge of being exhausted and cannot serve tourists effectively. Consequently, this has led to a decline in Himachal Pradesh's ranking concerning the influx of tourists, both from within the country and abroad, in comparison to other states. Thus, there is a need for formulating strategies that would help tap the vast potential of rural tourism available within the state. The rural population would be provided with an alternate source of earning, reducing the dependence on the turbulent and volatile agriculture sector. This would check the widespread migration of people from rural areas to urban areas in search of jobs by ensuring savings and investment habits among the rural people and thereby necessitating financial instruments. It is in line with the government's objective of financial inclusion. Apart from the financial benefits, rural tourism would ensure social inclusion as well. Welcoming tourists to their rural roots allows people to showcase Himachal Pradesh's authentic splendor – encompassing customs, traditions, history, folklore, and genuine Himachali cuisine. This garners global attention and offers firsthand experiential learning. Such escalated experience would be fruitful for the local population in Himachal as well. The book reflects on the development of rural tourism in the Himalayan foothills. Various topics related to the tourism industry are assessed from different perspectives and summarized in nine chapters.

The first chapter highlights the economic importance of tourism by analyzing the economic contribution of the travel and tourism industry to global GDP and employment levels. It discusses challenges for the tourism industry at the global level and the strategic solution to these challenges. The second chapter presents an overview of the tourism industry in India and Himachal Pradesh. First, it discusses the flow of tourist arrival and foreign exchange earnings earned by the country through the tourism sector. Then, it discusses the role and objectives of

national and international agencies working for the development of the tourism sector in India. It also covers the agencies working for tourism development in the state and provides a critical assessment of the tourism policy followed by the state.

The third chapter presents the research design of the study. It provides a description and justification of the methodological approach and methods used in the study. It describes the research framework of the study, research objectives, and hypothesis. The forth chapter presents an overview of rural Himachal Pradesh. It talks about the present status and trends in rural tourism in the state. It evaluates the consequences of rural tourism on the socio-economic structure of Himachal Pradesh. It shows how rural tourism can be a game-changer in developing rural India by tackling the socio-economic challenges prevailing in society. The fifth chapter examines the demand of tourists and their perception of the success of rural tourism in Himachal Pradesh. It highlights the characteristics and knowledge of the demand-side in the tourism industry that will be crucial for determining the success of tourism products in the state. In the sixth chapter, the opinions of other stakeholders, i.e., hoteliers, restaurant owners, travel agents/tour operators, homestay owners, photographers, tourist guides, and adventurers, are discussed to shed light on their perception of the development of the tourism industry in the state. It helps practitioners in identifying the knowledge gap, if any, that exists between stakeholders and tourists. In the seventh chapter, the opinions of all the stakeholders are integrated to form a flexible strategic crystal for the development of the rural tourism industry. Chapter eight delves into significant Himalayan vulnerability to global climate shifts, addressing key challenges. As the mountainous ecosystems are most fragile, understanding the preponderance of issues related to climate change and tourism can help in fostering sustainability. In the ninth chapter, the key takeaway from the analysis and discussion done throughout the study are summarized to form the conclusion for the study. Aligning with contemporary trends, the stemming key suggestions unfold the integration and amelioration of tourism industry on sustainability ladder. I believe this book will be helpful in drawing global attention from policy-makers and other stakeholders.

New Delhi, India Suneel Kumar

Acknowledgments

I wish to express my profound gratitude to my Ph.D. supervisor, Prof. Kulwant Singh Pathania,who dedicated more than 30 years of service to the Department of Commerce at Himachal Pradesh University, Shimla. His guidance has been invaluable in shaping this research. I am equally indebted to Prof. Madan Lal, a distinguished figure in International Business and Marketing, from the Department of Commerce at Delhi School of Economics, University of Delhi. His consistent guidance and unwavering support have been instrumental throughout this journey. I extend my heartfelt appreciation to the Principal, esteemed colleagues, dedicated administrative officers, and the diligent library staff of Shaheed Bhagat Singh College (University of Delhi). Their constant encouragement has been a driving force. Special recognition goes to Robert Doe, Aaron Schiller, Henery Rodgers, Zachary Romano, and Herbert Moses, whose unwavering assistance played a pivotal role in bringing this book to completion. I am grateful to Springer Nature for affording me this invaluable opportunity to contribute. My sincere thanks also extend to Dr. Ram Lal, Dr. Anisha, Dr. Ashikesh, Shekhar, Neha Kaushal, Upasana Devi, Anita Kumari, and Babita Bansal, whose support in data collection was indispensable for the execution of this research. I am indebted to Dr. Geetanjali Sageena for her significant contributions towards the culmination of this research endeavor.

My heartfelt appreciation goes to my parents and family members for fostering an environment of support and productivity. Their encouragement and assistance during various phases of this work have been instrumental. I am deeply grateful to my better half, Dr. Richa, and my daughters, Shaarvi Kashyap & Nirvi Kashyap. Their unwavering support during challenging pandemic circumstances made this undertaking achievable. Above all, I acknowledge the guiding presence of the almighty God, whose unseen hand has been a constant source of direction.

I extend my sincere thanks to the Indian Council for Social Science Research (ICSSR) and the Ministry of Education, Government of India, for their financial support in this research endeavor under file number IMPRESS/P2321/308/2018-19/ICSSR.

New Delhi, India Suneel Kumar

Contents

List of Figures

List of Tables

About the Author

Suneel Kumar is a Professor at Shaheed Bhagat Singh College, Department of Commerce, University of Delhi (India). He has done his Ph.D. in tourism marketing and tourist behavior from the Department of Commerce, Himachal Pradesh University, Shimla (India). He has about 14 years of experience in teaching undergraduate, postgraduate, and research students in Delhi University. His area of research is tourism and hospitality management, destination branding and marketing, sustainable rural tourism, and niche tourism marketing. He has more than 20 research papers published in his name in several international and national level journals. He has presented more than two dozen papers in international and national conferences, workshops, and symposiums and has more than six chapters published in edited books. He was an academic council member, University of Delhi, from 2015 to 2019.

Introduction

1

1.1 Sustainable Rural Development in India

Rural area or popularly known as the village is classified as an area having low population density. As per a report published by Dhanlaxmi Bank (2010), the National Sample Survey Organisation (NSSO) defines "rural" as follows:

- An area with a population density of up to 400 per square kilometers.
- Villages with clear surveyed boundaries but no municipal board.
- A minimum of 75% of male working population involved in agriculture and allied activities.

As per the census of India (2011) report, it was estimated that about 89.9% of the population still resides in rural areas. The nation is home to huge assortment of villages crossing various districts, each having its own arrangement of customs and legacy. This suggests that not only India is home to the huge landmass that is classified as a rural area, but also it encompasses more than 70 crores people. India's demography is augmented by its geographic variety. From the freezing, snow-covered pinnacles of the Himalayas in the north to the sandy shorelines of the southern landmass, from the verdant slopes of the north-east to the burning desert in the west, the nation is home to a huge assortment of scenes. This huge amount highlights opportunities present in the rural areas as well as the dire need for rural development. Rural development means raising the standard of living in rural areas by developing required infrastructure such as telecommunication and road networks, hospitals, schools, etc. and providing markets with a variety of goods for consumption. It additionally implies raising the per capita income level of the rural people so that they can likewise appreciate the luxuries in life. Since India has been traditionally a rural economy (as a large of India even today resides in rural areas), it has been pretty challenging for the country to promote a balanced rural development in every state. According to the reports by the World Bank, there still exists 65% of the

S. Kumar, *Sustainable Rural Tourism in Himalayan Foothills*,
https://doi.org/10.1007/978-3-031-40098-8_1

population in India (2018–2019) which is living in rural areas. Although the trend is diminishing, there is still a long way to go before urbanization assumes control over the entire country. Since there is a humongous presence of rural areas in India, therefore, without rural development, the aim of a 5 trillion USD economy cannot be accomplished. Although there is rural development in some states, few poor states endure to an extent that there exists a lot of divide between urban and rural development overall.

Cultural and natural assets are key factors in advancing reasonable improvement of tourism industry (del Río-Rama et al. 2020). In response to globalization, the significance of the sustainable development idea in tackling the issues of tourism industry framework advancement is gaining attention. In view of this, it is important to evaluate rural tourism potential as a holistic idea and important to incorporate all elements that might influence its future improvement. An arrangement involving adherence to three paradigms, i.e., natural and social variables with economic standards, becomes progressively crucial. The concentrated quest for new methodologies that guarantee expansive changes at the worldwide level, however not absolved from empty way of talking what's more purposeful equivocalness, prompted the emerging of the idea of sustainability that turned to provide a better vision at a common stage. A plethora of studies have been conducted highlighting the importance of rural development, and they all suggest setting up of industries for the same. But, still, the desired target achievement is far from reality. The government has not been able to find a strategic solution for rural development. Since industry development is a well-thought process, it generally takes much time before benefits begin to emerge out of industrial development. In addition, industrial development has its own repercussions as well. Similarly, till now there is no single industry that could be established everywhere to promote rural development. Rural tourism as an agribusiness enterprise has the potential to enhance rural communities' socioeconomic growth and provides several advantages for enhancing their culture, way of life, and technological approach. However, the venture's viability depends on community involvement. Because of all these issues, rural development is yet to be achieved completely in India.

1.1.1 Critical Assessment of Steps to Promote Rural Development in India

Rural urbanization is causing a global economic, social, and cultural evolution of rural urbanization to the quick and constant improvements in urbanization. Rural areas' economic, social, and environmental aspects are all included in rural tourism. It has particular implications on the environment and on economic growth and is strongly connected to people, space, and products. In addition to being a method to support sustainable rural development in developing nations, urbanization is a universal trend that must be addressed. The government of India has progression of ways to fulfil its objective of rural development. The major concern for the government lies in the fact that in the absence of rural development, a major share of the population has started to migrate from rural regions to urban regions in search of employment and a better standard of living. This has created an unnecessary pressure on the urban regions to accommodate people more than its carrying capacity. Literacy level has often been linked with the level of development and standard of living in a region. Since the rural regions are marred with low literacy among its population, the government has taken a series of steps to enhance the literacy level among rural India. Due to its persistent efforts, the literacy rate in rural India has risen from 36% in 1991 to 69% in 2011 as per the census of India (2011). Despite the fact that there is a quick expansion of rural literacy, rural development is yet to take its flight. The government has taken several steps for rural development. These steps are critically analyzed as follows:

- Financial inclusion: Financial inclusion is defined as the level to which individuals and businesses have access to useful and affordable financial products and services that meet their needs that are delivered responsibly and sustainably. Financial inclusion is defined as the availability and equality of opportunities to access financial services. Financial inclusion strives to remove the barriers that exclude people from participating in the financial sector and utilizing these services to improve their lives. It is also called inclusive finance. The government has taken a series of steps to promote financial inclusion in rural India. The Pradhan Mantri Jan Dhan Yojana (PMJDY) aims to persuade people to open their bank accounts in nationalized, public, regional, or private sector banks to enable them to use financial services such as obtaining loans, using credit facilities, direct benefit transfer schemes, and use of digital payment mechanism. The banks were successful in persuading people in rural India to open their no-frills accounts, yet, these records by and large remain exceptionally dormant. This is because people don't have any reason to use these bank accounts, due to their limited exposure to opportunities for the same. This is a primary reason that people are not only reluctant to open the account but also reluctant for their use. This is one of the major obstacles in the mission of making India financially inclusive. Similar schemes like expanding social security net and establishment of Micro Units Development and Refinancing Agency bank have also failed to achieve the desired benefits for financial inclusion.
- Social inclusion: Social inclusion is characterized as the way toward improving the terms on which individuals and groups take part in society improving the ability, opportunity, and dignity of those disadvantaged based on their identity. Social inclusion is the process by which efforts are created to ensure equal opportunities—that is, everyone, regardless of their background, can achieve their full potential in life. Such efforts include policies and actions that promote equal access to public services as well as enable citizen's participation in the decision-making processes that affect their lives. The government has its schemes and program to make people aware about the different culture and traditions of various ethnic groups living in India.
- Alleviating the poor through poverty alleviation schemes- Despite the fact that poverty is a multifaceted problem, economic factors based on income and con-

sumption are frequently used to quantify poverty levels. Poverty alleviation has been agenda of government since the last few decades. Significant improvement has been apparent in achieving the targets over the years. The poverty rate in India has declined to below 10% in the year 2019 as reported by the National Statistics Commission. However, the report also suggests that rural poverty has increased in India after a few decades. The same report highlights the rural poverty rate at 29.6%. This highlights that impact of poverty alleviation schemes has been felt more in urban areas than in rural areas. Schemes such as Mahatma Gandhi National Rural Employment Guarantee Act, Integrated Rural Development Programme, Annapurna Yojana, Jawahar Gram Samridhi Yojana, etc. have failed to accomplish the ideal outcomes for rural development through poverty alleviation in India. Although the numbers might have reduced in recent years, the status of people has not improved much. They have successfully managed to beat the poverty line, but there is not much progress in the living standard of people. Leave aside the luxuries; they are yet to enjoy the benefits of the several basic commodities as well.

- Employment generation through skill development: Schemes such as "Stand-Up India" aims to create an atmosphere of subaltern entrepreneurship. Since employment generation is a massive task in the rural region, and the absence of which results in migration, the government has started to promote skill-based entrepreneurship among rural people. Thus, individuals who were once running in search of seeking a job are now working to hire people to do the jobs. From job seekers, they are being built as job givers. However, rural entrepreneurship is yet to achieve significant heights. There is a lack of industries in rural regions that require skills for jobs. Since a lot of industries are manufacturing based, hence, there is a lack of development of diverse skills in an individual.
- Setting up manufacturing industries: Rural development through setting up of manufacturing industry has long been utilized as strategies by the government. The government has dedicated 5 years of plans for the same objective. Setting up of special economic zones, export processing zones, and industrial areas has been an approach for doing the same. However, industrial development has its own pros and cons. Rapid industrialization has taken a toll on the environment of the region. Additionally, not every region could support due to the lack of resources and infrastructure. Further, the government has taken too much time to identify one suitable industry that could be set up in all the rural areas. This has resulted in unbalanced regional development. Some regions have prospered because the industry was able to survive thereon. The government has not given the service sector its due significance in the improvement of rural regions till now.
- Making industries more inclusive: It is seen that the benefits of industrial development, be it secondary sector or tertiary sector, are enjoyed by the few. The industries were originally more exclusive in nature, i.e., the access to participate in the industry was limited only to a few. However, recent studies in strategic management have highlighted the concept of the bottom of the pyramid. As per this concept, a lot of business opportunities can be ascertained within the people belonging to the economically weaker section of the society. The government is

trying to incorporate this principle while making arrangements for industrial development. With emphasis on poorer sections of the society, visions and missions for the industry are being developed to promote inclusiveness in the industry.

In a nutshell, it can be summarized into few observations. First, rural development has been on focus of government, and schemes and policies were developed and are being developed for the same. Second, the desired results for rural development have not yet been achieved. Third, poverty alleviation programs are effective, but, recent trends in rural regions are not encouraging for the same. Mere upliftment from the poverty line is not benefitting much for the cause of rural development. Fourth, the manufacturing sector-led development has not been able to provide a significant support to the rural people. Subsequently, there is a need to shift the horizon to the service industry. The service sector comprises industries such as the financial sector which includes banking, insurance and allied services, communication services, medical sector, transportation sector, tourism and hospitality sector, etc. The tourism and hospitality sector has gained the attention of policymakers due to its tremendous potential in the socioeconomic development of the region. Thus, this study focuses on using the development of the tourism industry as a tool for transforming rural India. For this purpose, rural tourism which is a form of niche tourism has been considered for the present study.

1.2 Introduction to Tourism

Tourism is defined as the movement of an individual traveling to and staying in places outside his usual environment for not more than 1 consecutive year for leisure, business, and other purposes. According to Erik Cohen, "A tourist is a voluntary/ temporary traveler. He is traveling in the expectation of pleasure from the novelty & change experienced on a relatively long & non-recurrent round trip." Tourism is not a new concept. Since long, people have been visiting from one country to another on an expedition or for spreading or gaining knowledge about different cultures, religions, or economic systems. The nineteenth Century Dictionary defines "tourist" as "a person who travels for pleasure of traveling, out of curiosity & because he has nothing better to do." Tourism is a diverse concept. It has economic, sociological, and ecological aspects. Tourism sustainability entails the promotion of sustainable tourism tactics, such as creative firm-territory relationships, value cocreation, and financially sound, long-lasting company models and operations.

1.2.1 Tourism Typology

Tourism exists in various forms; a tourist can be classified based on its origin and destination and based on the purpose of its visit. The classification of tourists is of utmost importance; it not only helps in knowing about the reasons for which a destination is popular, but it also helps in identifying what are the opportunities to tap. If a destination can attract only a certain type of tourists, there is a chance that a certain natural threat or any policy change might render a serious threat to the tourism industry at that place. Further, the classification also helps in distinguishing the economic contribution of tourists from different categories and also helps in strategizing as per requirement. Integration of tourism for development might vary enormously across various geographical locations. Tourism typologies applicable to tourism reflect a continuum synergism required to achieve the sustainable tourism. The different types of tourism are as follows:

- Domestic tourism: When a person who is an inhabitant of a nation visits a place located outside its usual residential premises, but, located within the same country, then this type of tourism is known as domestic tourism. So when a resident of Maharashtra, India, traveling to another state Kerala located in India will be termed as domestic tourists. Domestic tourism is essential as it encompasses the major chunk of tourists visiting any place. No destination can survive economically without domestic tourists. They help in setting up and maintain the infrastructure which further could be developed to attract foreign tourists.
- Outbound tourism: When a person who is a resident of a country makes a visit to a destination located outside the country of residence, it is termed as outbound tourism. A journey by a resident of India visiting the United Kingdom will be termed as outbound tourism. A country needs to keep statistics of its outbound tourist for its economic gains. It helps in identifying the reasons for which tourists prefer to go outside the country and can lead to the development of niche tourism products that are nonexistent in the native country. Likewise, outbound tourism leads to leakage of foreign exchange, so the country should try to develop the industry as such that it limits the need to travel outside.
- Inbound tourism: When a person who is a resident of another country and makes a visit to a destination located inside our country, it is termed as inbound tourism. So, a resident of China visiting India will be termed as inbound tourists from the viewpoint of India. A common term for inbound tourists is "foreign tourists." Inbound tourism brings economic gains to a nation in terms of foreign exchange earnings. They also bring social and technological advancement to the destination which they visit. Governments put a lot of emphasis on the satisfaction of inbound tourists as they create goodwill for the visited nation in their native country.
- Leisure tourism: Perhaps the most popular types of tourism comprises activities such as nature tours, visit to cultural heritage, and sites for the purpose of recreation and enjoyment. People seldom visit theme parks and leisure programs to destress themselves from the hectic schedule. As per a report by the US Travel Association, the leisure tourism industry alone itself is of US$ 761.7 billion and accounts for 6.4 million jobs directly in the year 2018. Such is the share of leisure tourism in developed countries that almost 80% of trips undertaken by tourists are for leisure tourism purposes.

- Religious tourism: The world comprises of about 12 major religions and approximately 4300 total religions. Tours and trips underwent for meeting religious purposes are one of the most consumed tourism typologies particularly in developing and underdeveloped countries. People across the globe move cities, countries, or even continents for the purpose of religious tourism. The UNWTO estimates that tourists take 600 million national and international religious trips per year. It also estimates that 300 million tourists visit the world's significant religious sites each year.
- Pilgrimage tourism: One of the subsets of religious tourism is the pilgrimage tourism. In this type, people visit places that have the utmost importance in their religion. Examples of pilgrimage tourism include Haj (performed by Muslims throughout the world) and Kumbha Mela practiced in India by Hindus. The significant distinction between a pilgrimage tourist and a religious tourist is the extent of faith. Pilgrimage tourism is consumed at the utmost level of faith of a devoted individual. As per a report by Alliance of Religions and Conservation in 2011, it was estimated that pilgrimage tourists would reach the mark of 155 million tourists in the year 2012 (ARC, 2011).
- Wellness tourism: It is defined as travel to promote health and well-being through physical, psychological, or spiritual activities (Dimon, 2014). The objective of wellness tourism is to improve and maintain health and quality of life (Kumar et al. 2018). As per the Global Wellness Tourism Market report by Technavio, it is assessed that the wellness tourism market will witness a 10% growth rate annually over the years 2017–2021. Moreover, Asia-Pacific countries will be the major hotspot for wellness tourism for the upcoming years.
- Spiritual tourism: Spiritual tourism is to venture out to discover purpose and meaning in your life. It elevates your physical, mental, and emotional energies. It develops, maintains, and improves your body, mind, and spirit. In a nutshell, it connects your body, mind, and soul. Hence, spiritual tourism can be defined as traveling on pilgrimage to acquire enlightenment, entertainment, and education to leave your footprints and take your memories back to your home by expressing your gratitude to God. Spiritual tourism includes following spirituality in daily routine, consuming sattvic food, and changing daily habits. The aim of spiritual tourism is to live in an ambiance of purification which leads to spiritual transformation.
- Medical tourism: Medical tourism is a term comprising of tours and travel activities by a person for undertaking a medical treatment. Various nations in South Asia and Southeast Asia have become a prominent hub for medical treatment for patients across the globe. In addition, high-class patients tend to go to the USA for treatment at super-specialized hospitals. The major reason for becoming a medical hub is the treatment by super-skilled doctors with the use of high-technology equipment and at a very minimal cost. Kumar et al. (2018) highlighted that the medical cost treatment in India is much lower than the cost for the same treatment in the USA or other developed countries. There are various benefits that patients get due to medical tourism. It includes savings in treatment,

less waitlist, high-quality treatment with world-class facilities, customer care, and also availing travel opportunities.

- Ecotourism: In the 1980s, at the beginning of sustainable development, ecotourism was developed as a means of directing tourism income toward development and conservation. Ecotourism is now defined as "responsible travel to natural areas that conserves the environment, sustains the well-being of the local people, and involves interpretation and education" (TIES, 2015). Ecotourism is a form of tourism involving visiting fragile, pristine, and relatively undisturbed natural areas, intended as a low-impact and often small-scale alternative to standard commercial mass tourism.
- Dark tourist: Dark tourism is understood as a travel motive to experience death or near-death experiences. It is also known as "Thanatourism" and is a growing niche tourism segment. It is also a way to remember those who have lost their lives to a natural or man-made disaster. Major famous sites for dark tourism include Ground Zero in the USA, concentration camps in Auschwitz, Museums in Hiroshima and Nagasaki, etc. The major reason for the growth of the dark tourism segment is the fascination of youth to undertake something beyond ordinary. To attract followers on their social media platform, they, in search of something new, have made it a popular segment.
- Adventure tourism: A significantly popular form of tourism among the youth worldwide, adventure tourism has its own share in attracting tourists. To experience the adrenaline rushing through their veins, youth want to engage in activities that provide fun, joy, and excitement. It includes river rafting, paragliding, trekking and hiking, etc.
- Food tourism: It is also known as gastronomical or culinary tourism. It is characterized as tourism for the sole purpose of exploring the food items. It is about what is unique, authentic, and memorable about the delicious stories a destination has to tell. This includes farmers, cheesemongers, fishermen, brewers, winemakers, and everyone in between (Kumar et al. 2018). The top food tourism activities include dining out at a gourmet restaurant, eating street and local food, and visiting a famous restaurant and bar. As per world food travel association report on 2020 food travel trends, 53% of all travelers are culinary travelers. The same report also highlights that a culinary traveler spends 25% more than a normal tourist at a destination. This proposes that more economic gains could be generated by developing the food tourism industry.
- Virtual tourism: A virtual tour is a simulation of an existing location, usually composed of a sequence of videos or still images. It may also utilize other multimedia elements such as sound effects, music, narration, and text. It includes the use of artificial intelligence, augmented reality, and virtual reality in creating tourism products or in using these to serve the tourists. The use of virtual reality increases tourist's satisfaction.
- Wedding tourism: It is defined as traveling outside the residence of both bride and groom for the purpose of a wedding. Another term used more commonly for the same is the destination wedding. Because of a rise in disposable income of the middle class, as well as the promotion of destination wedding in movies and

TV shows, wedding tourism is on roll in India. Owing to the emergence of wedding planners and supportive infrastructure, wedding tourism is one of the top niche tourism products in the country. According to reports by FICCI, the wedding tourism market in India is of about Rs. 23,438 crores and is expected to grow to Rs 45,000 crores by the year 2020. Cities like Jaipur, Udaipur, and Jodhpur in Rajasthan, Goa, Himachal Pradesh, and Jammu and Kashmir are among the top choices for destination weddings in India. Aamby valley city and Ramoji film city are also few spots that have been well established as hotspots for wedding tourism in India.

- Halal tourism: Halal tourism is a form of tourism where activities are undertaken by keeping in mind the rituals, customs, and traditions of Islam. It includes a prohibition on serving alcohol in hotels during the stay, serving only halal food, developing Muslim-friendly mobile applications for navigation and other purposes, and developing halal holiday packages and health care facilities and services. According to the State of the Global Islamic Economy report, produced by Thomson Reuters in collaboration with Dinar Standard, the global Muslim travel market is expected to be worth $238 billion in 2019 and represents 13% of global expenditure. The leading Halal trends in the tourism industry currently include the need for halal advisory, rising female voice in trip planning, hotel sectors becoming more sensitive to Muslim necessities, deeper training immersion with VR Umrah, and consolidation of Muslim travel space.
- MICE tourism: MICE is an acronym for meeting, incentive (travel), convention, event/exhibition. It includes all those tours and trips that are undertaken for the abovementioned purposes. It includes corporate conferences, trips planned for training employees, academic conferences hosted by national and international organizations (such as CoP, BRICS summit), and cultural and sporting events organized by a nation (such as the Olympics, Cricket World Cup, or other comparative events).

1.3 Rural Tourism

Rural tourism is described as the tourist visit to a rural area for active participation in a rural lifestyle. It is described as a travel activity in the premise close to the rural region. It includes consuming nature-based activities such as hunting and fishing, agriculture produce, walking in the jungle or desert safari, living local lifestyle, adopting local customs and traditions, consuming local cuisines, using local handicrafts, enjoying local festivals and folklores, and living and residing in local ambiance rather than in a hotel or a resort. It is basically traveling to get close to nature in rural premises. Rural tourism is now being practiced as a fully developed tourism product rather than just being a niche. Since there is a lot of scope of rurality in developed as well as developing countries, rural tourism has witnessed growth in attracting tourists worldwide. This type of tourism is widely practiced in the countries of Africa and Europe which widely attracts tourists from all corners of the world. The present study tries to establish a link between developing rural tourism and promoting rural development. Not much of thought has been given in considering rural tourism to be an industry for ensuring rural development. Since rurality is present around all the states of India, therefore, there is a high chance that rural tourism can be developed since each state in India is enriched with its own culture, tradition, scenery, food, and lifestyle. The only significant hindrance that exists in the development of rural tourism is the issue of sustainability. Since resources are scarce and in rural areas where means to acquire these resources are also limited, there is a chance of crowding out of resources for the local population. This will affect the quality of life in the rural area. Therefore, while developing rural tourism, it should be noted that practices followed are as per the principle of sustainability. So, sustainable rural tourism can be developed to promote rural development in India.

1.4 Sustainable Rural Tourism

As a result of a high impact on development, there is a need for a sustainable approach in tourism, all the more significantly in nations that are on the path of developing and are experiencing a high growth rate, which is imperative. Tourism is one of the human activities that have caused major damage to the environment. The United Nations World Tourism Organization which is a leading agency in the promotion and development of tourism dedicated the year 2017 to be the "International Year of Sustainable Tourism for Development." As per UNWTO, sustainable tourism is defined as "Tourism that takes full account of its current and future economic, social and environmental impacts, addressing the needs of visitors, the industry, the environment and the host community." Buckley (2002), for instance, defines sustainable tourism as "tourism at any scale with practical and proactive design, engineering, and management to reduce environmental impacts." Lane (1994, p. 2) explicitly referred to sustainable tourism as "providing satisfying jobs without dominating the local economy or abusing the natural environment." Tourism is said to be sustainable when its operation and development include the participation of the local population, protection of the total environment, fair economic return to the industry, and host community, as well as mutual respect for all the stakeholders involved. Rural tourism is described as the tourist visit to a rural area for active participation in a rural lifestyle. Rural tourism which started as a sub-niche of ecotourism is now being considered as a major tourism product for its socioeconomic benefits. An increase in employment, development of infrastructure, a rising standard of living, exchange of ideas, knowledge, customs across communities are some of the benefits provided by the tourism industry. Recent studies have highlighted that certain types of rural tourism are not sustainable in nature. They accomplish more damage to nature and surroundings than provide benefits. Therefore, there is a need to introduce certain measures that help in the development of rural tourism sustainably. One must not confuse sustainability in tourism by restricting tourism to protect the environment. Strategies must be formulated in a manner that the satisfaction of tourists is not reduced and damage to the environment is minimal.

1.4.1 Gender Inclusion in Sustainable Development

An integrated approach to sustainable rural tourism must be inclusive of indispensable gender perspectives and critical focuses ought to be considered across three dimensions, i.e., social, economic, and environmental. The infusion of cross-cutting gender equality shall aim to foster and consolidate rural sustainable tourism development.

The travel and tourism industry at local level should be planned with regard to the power relations created by access to assets and of the adverse consequences on the capacity advancement of both males and females. Concurring with the UN affirmation of 2017 as "the International Year of Sustainable Tourism for Development," the travel industry area will actually want to show its eagerness to ensure strengthening gender inequalities. The significant focuses target must be as follows:

- For the effective implementation and monitoring of the sustainability of tourism, it is essential to make an investigation of the gender interests in various areas connected to the travel industry like agronomy, administrations, multilateral cooperations, and development just as a gender analysis identified with the travel industry in private sector.
- Responsible consumption apart from independent decision-making is yet additionally needed to expand the security of females involved as a part of travel industry objections and for females who travel by their own.
- As a part of fundamental financial planning, legislatures need to execute the travel industry techniques that become true piece of the country's travel industry. The initial step is to comprehend the real factors of the participation of women who are engaged with the travel industry. The creation and execution of the travel industry strategies obviously require corporate, social, and ecological obligations.
- A need in direction of decent work requires accentuation on salary gaps holes, sexual harassments by partners and sightseers, and promoting female participations.

1.5 Economic Contribution of Tourism Industry: An Appraisal

The tourism sector has assumed a pivotal role in economic development through job creation throughout the world (Zolfani et al. 2015; Mir 2014). Not only the impact is apparent in the form of the number of jobs created, but it can also be assessed through contribution to GDP, visitor spending, contribution to capital investment, and other similar indicators. Table 1.1 demonstrates the key contribution of the tourism industry in the global economy in the years 2018–2019.

The key highlights from Table 1.1 are:

- Talking about the direct contribution of the travel and tourism industry to global GDP, the contribution which was USD 2750.7 billion (3.2% of GDP) in 2018 is

Table 1.1 Contribution of tourism industry in the world economy

World	Billion USD	% of total	Growth	Billion USD	% of total	Growth
Direct contribution to GDP	2750.7	3.2	3.6	4065.0	3.5	3.6
Total contribution to GDP	8811.0	10.4	3.6	13085.7	11.5	3.7
Direct contribution to employment	122,891	3.8	2.2	154,060	4.3	2.1
Total contribution to employment	318,811	10.0	2.9	420,659	11.7	2.5
Visitor exports	1643.2	6.5	4.0	2483.9	7.2	3.8
Domestic spending	4060.1	4.8	3.5	6031.9	5.3	3.7
Leisure spending	4475.3	2.5	3.8	6780.7	2.8	3.9
Business spending	1228.0	0.7	3.0	1735.1	0.7	3.2
Capital investment	940.9	4.4	4.4	1489.5	5.0	4.2

Source: World Tourism & Travel Council Report, 2019

Table 1.2 District-wise area and population of Himachal Pradesh (2011 census)

Sr. No.	District	Area (sq. Km.)	Population
1	Lahaul and Spiti	13,835	31,564
2	Chamba	6528	519,080
3	Kinnaur	6401	84,121
4	Kangra	5739	1,510,075
5	Kullu	5503	437,903
6	Shimla	5131	814,010
7	Mandi	3950	999,777
8	Sirmaur	2825	529,855
9	Solan	1936	580,320
10	Una	1540	521,173
11	Bilaspur	1167	381,956
12	Hamirpur	1118	454,768
	Total	**55,673**	**6,864,602**

Source: Himachal Pradesh Statistics Report, 2012

forecasted to increase by 3.6% to USD 2849.2 billion in 2019. The main reason for such an increase is the expanding trend in revenue generated by the hospitality sector including hotels, travel agents, aviation sector, etc. as per the estimates; the direct contribution is supposed to increase by 3.6% p.a to USD 4065 billion (3.5% of GDP) by the year 2029.

- The total contribution of travel and tourism to GDP which includes direct as well as indirect contribution was recorded to be USD 8811 billion in 2018 (10.4% of GDP) and is expected to rise by 3.6% to USD 9126.7 billion (10.4% of GDP) in 2019. The same is expected to rise by 3.7% pa to USD 13,085.7 billion by 2029 (11.5% of GDP).
- Expenditure made by foreign tourists acts as a key component of the direct contribution of travel and tourism. It is also the main source of earning foreign

exchange reserves for a country. In 2018, the world generated USD 1643.2 billion in visitor exports. In 2019, it is estimated that the number of international tourists will be approximately 1,484,910,000, and this will lead to an increase of 4% in expenditure made by them. By 2029, international tourist arrivals are forecast to total 2,196,090,000, generating expenditure of USD 2483.9 billion, an increase of 3.8% pa.

In light of the facts and figures mentioned in Table 1.2, the tourism industry can be attributed as a valuable agent for the development of the economy. The role of the tourism industry for the economy can be understood through the following points:

- Tourism development creates employment: Tourism creates direct and indirect employment. Direct employment is due to the development of the tourism industry such as guides and photographers, and indirect employment is the result of setting up several other industries due to the development of the tourism industry. Reports of WTTC suggest that around 4% of total jobs generated in the year 2018 are the result of the contribution of tourism industry development. Travel and tourism generated about 122,891,000 jobs directly in 2018 (3.8% of total employment), and this conjectures to develop by 2.2% in 2019 to 125,595,000 (3.9% of total employment). This includes employment by hotels, travel agents, airlines, and other passenger transportation services (excluding commuter services). It also includes the activities of the restaurant and leisure industries directly supported by tourists. By 2029, travel and tourism will account for 154,060,000 jobs directly, an increase of 2.1% pa over the next 10 years (World Tourism and Travel Council, 2018).
- It promotes infrastructure development and maintenance: Development of tourism industry helps in setting up of infrastructure such as road, connectivity, communication, sanitation, electricity connection, etc. This could be termed as an externality originating due to the tourism industry. The infrastructure development leads to the rise in the living standards of the people in the region. According to the reports, travel and tourism is expected to have attracted capital investment of USD 940.9 billion in 2018. This is expected to rise by 4.4% in 2019 and rise by 4.2% p.a. over the next 10 years to USD 1489.5 billion in 2029. Travel and tourism's share of total national investment will ascend from 4.4% in 2019 to 5.0% in 2029 (World Tourism and Travel Council, 2018).
- It creates an alternative source of earnings: Tourism is beneficial in rural regions in a country as it provides them an alternative to the agriculture and primary industry for earning their livelihood. As this industry does not require much technical knowledge or a professional degree, the tourism industry offers opportunities to such people who need employment.
- Contributes to foreign exchange earnings: Depleting foreign exchange reserves has always been a worrisome problem for developing nations. By attracting foreign or international tourists, a country can create a means to earn foreign exchange reserves.

- It leads to the exchange of products, services, and technologies: When people visit foreign countries, they consume products and services which they are not familiar with. As a result, they bring these products with them back to their home. Similarly, when tourists visit any other nation, he/she introduces products to the destination country. Interestingly, Samosa, a very popular snack in India, is an example of a product first introduced to them by foreign Persian tourists.
- It helps in building an image for a nation: Tourism is a way to fabricate a connection between nations. Hospitality experienced by tourists in a country helps in building a favorable image of the nation. If tourists are happy with a country, they will feel more connected to a country, which will also help in building friendship among nations. It will also be visible in the pro-country policies developed by the government.

1.6 Challenges Observed in Development of Tourism Industry

Although the tourism industry is beneficial economically and socially for a country to develop, this industry is under a lot of stress over a recent period. When an industry is running smoothly, it encounters several issues in its working which might act as a catalyst in its fall. Since tourism is one of the major activities contributing to the GDP of several states, it is imperative to understand the challenges surrounding it as well.

- Over tourism at certain destinations: Several destinations that are popular across the globe have started to suffer from the problem of over-tourism. The concept of carrying capacity, which is of utmost importance in tourism, is highly neglected at these destinations. Over-tourism has not only decreased the attractiveness of the destination, but it has also started to create destination hate at some level or other.
- Socio-social effects related with expanded interest and appearance incorporate hindering the day-to-day exercises of neighborhood occupants because of congestion and clog at attractions, out in the open spaces, and on open vehicle. Development in guest numbers past the normal conveying limit of objections can likewise include pressure existing framework and administrations, in this manner adversely affecting the experience for local people and guests the same. In metropolitan regions, the spread of sightseers past conventional the travel industry "zones" into some time ago private areas can prompt the estrangement of neighborhood networks, because of improper conduct of vacationers, expanded clamor contamination, and person on foot and vehicular traffic, as well as the corruption of local area explicit framework and offices. Different effects related with expanded tourists can incorporate a deficiency of genuineness and social personality/legacy, for instance, an increment in cordiality, social and retail benefits for visitors, and rivaling and uprooting inhabitant-arranged administrations and offices. In addition, the course of adjusting and making changes in accordance

with addressing the issues, mentalities, and upsides of travelers can prompt the alteration of occasions, exercises or social locales, or the misfortune or debilitating of social practices and values.

- Assessment of gender incorporation: Identifying the reasons for discrimination that happen straightforwardly and by implication because of the advancement of the travel industry.
- Depleting natural reserves: A significant issue experienced at destinations that are attractive because of the presence of natural resources is their continuous depletion. Because of the consumption of these resources at a growing rate, there is very little chance of their replenishment. Several destinations have been lost or are on the verge of being lost due to the depletion of resources.
- Externalities of tourism: Tourism has certain undesirable effects on the society, ecologically and socially as well. These undesired impacts can be termed as externalities of tourism industry. Be its depletion in the social values due to behavior exhibited by the tourists or the damage and destruction of the environment due to it, tourism has often been criticized and discouraged because of it. In the mid- to late 1980s, there began a reassessment of the role of tourism, and the focus of research shifted toward the potential negative impacts of the tourism industry. For long, tourism was considered to be a nontoxic pleasure-giving activity, but the rise of environmentalism and "green consciousness" challenged this perception, and it was realized that tourism was not a "smokeless industry."
- Under-exploring various destinations: A major challenge in front of the tourism industry is holistic development, i.e., development of the tourism industry at all the places that have the potential to be tourism destinations. Policymakers have regularly failed in this aspect. Certain destinations become a hotspot, where others are lost in the darkness of under-exploration.
- Negative image building for countries: If tourism can be used to build an image of a nation, cognitive dissonance experienced by the customers can often lead to damaging the reputation of a nation. Time and again, press releases highlighting advisories issued by the countries to their travelers while traveling to nations with crime rates against women are seen issued in newspapers and others. This tarnishes the image of a nation on a global platform.
- Too much dependence on tourism leads to repercussions: A famous investment quote states "do not put all your eggs in one basket." The equivalent is the situation with numerous governments depending solely on tourism for their economic development. The success of the tourism industry depends on several autonomous factors for which a strategy could not be developed or implemented. Therefore, if an economy is too dependent on tourism, and gets impacted by a factor, damage of which is beyond repair, then the whole economy gets pressured and leads to a collapse.
- Ecological effects from expanded use are commonly the consequence of consistent tourism development, critical pinnacles identifying with irregularity, or more unexpected, transient shocks, like the visit of an enormous voyage transport or the facilitating of significant occasions. Undeniable degrees of clog can prompt the disintegration of the travel industry-related framework, offices, and

nearby greenery and surpass the limit of help administrations to adapt to request, for example, transport foundation and water and waste executives (e.g., plastic, food, strong waste, water). Over-tourism can make mileage on assets like delicate normal and recorded destinations. Other potential effects incorporate expanded air, light, and clamor contamination and greenhouse gas outflows related with the vehicle of travelers and supplies to/from/around objections.

- Exhausting marketing mix: Marketing mix of tourism destinations is getting extensively exhaustive. Similar products are being offered without much innovation. Pricing at destinations is competitive without much change in the advertising and promotional strategies. Not much of celebrity endorsement is visible in promotions which could attract more tourists. New product development is hardly visible in the industry, and hence value addition to the industry has stagnated.
- The lack of seriousness in policymaking: Planning is a continuous process that needs to be continuously monitored, regulated, and followed up. Planning in tourism has been in general found to be rigid in nature. Planning is done without much vision, and missions are inflexible to such an extent that they leave very little scope for flexibility in innovation in plans. This casual approach has caused the underdevelopment of the tourism industry.

1.7 Tourism Industry Development: Corrective Measures

All the discussed challenges need to be addressed for the betterment of the tourism industry. Since the industry has tremendous potential to grow and cause benefit to mankind, several steps are needed to guarantee its endurance over the long haul.

- *Practicing Sustainability in Tourism*: To ensure that destinations continue to remain an attractive destination for tourism, it should follow the concept of sustainability. The number of tourists visiting a destination should be monitored continuously, there should be a regular check of the level of natural resources at these destinations, and a balance must be maintained between tourist arrival and carrying capacity of the destination. Planning should be such that it does not hamper the vision and mission of policymakers, protects the interest of every stakeholder, and does not reduce the satisfaction of the tourists.
- *Social and Gender Inequalities*: There is a pertinent need to rethink and realign the tourism development with mainstreaming of the promotion across social and gender-based equalities.
- *Development of New Tourism Products or Markets*: Tourism is an industry where supply creates its own demand. If new products could be developed and presented before tourists in attractive ways, they could pull tourists from overcrowded places and products. There has not been much advancement visible in tourism products in terms of innovativeness and inventions. New product development would also alter the marketing mix followed by the policymakers posi-

tively. They should focus on how to develop products from existing resources or create something new out of nothing. Development of new tourism products or markets can be done by conducting an assessment of the market about the resources present within the system and products that can be developed through these resources. Further, existing demand-supply analysis could be made and potential customers could also be identified for the future tourism products being developed.

- *Destination discontinuity*: Policymakers should follow the practice of destination discontinuity. As per this approach, destinations that have become exhausted or will soon turn into one need to be discontinued, and the flow of tourists should be diverted to destinations that have not been explored but have the potential to be developed alike. Along these lines, the economic loss resulting from discontinuing a flourishing destination could be minimized in earlier stages and fully mitigated in later stages of destination development. For this, new product development and identification of new destination are prerequisite.

1.7.1 Destination Discontinuity Model

The graphical representation for the model of destination discontinuity can be done as follows: on a graph, having a horizontal and vertical axis and on the horizontal axis, place period depicting the growth of destination over time. Similarly, on the vertical axis, it is the performance of the destination. The life cycle of a destination is in a way similar to that of a product. It, like a product, passes through different stages in its life course. When the destination starts reaching the maturity level, it began to lose interest among the tourists, and the footfall of tourists starts decreasing. This is the time when a destination must be discontinued, and efforts must be made for sustainable development of the destination while simultaneously looking for a new tourism destination (Fig. 1.1).

The graph highlights that when a destination (D1) reaches its saturation point, then there is a requirement to search for other destinations (D2) which can help to discontinue the prior destination or divert the tourist inflow. To make the new destination an alternate to the previous destination, it could be built up attractively through the destination branding process. The new destination ought to be sustained in a destination incubation wherein efforts should be done to ensure its survival. The new destination should also be ready in terms of infrastructure for serving the tourists. The theory proposes that, if all goes good, then the new destination will attract higher tourist inflow if there is an effective advertising and if more focus is given at the initial stage on local tourism than foreign tourist arrival.

- *Flexible strategic management*: The tourism industry is totally dependent on the environment surrounding it. Since the nature of the environment is dynamic, subsequently there is a need that plans developed for the tourism industry are also flexible in nature, i.e., they can assimilate new environmental factors, adjust

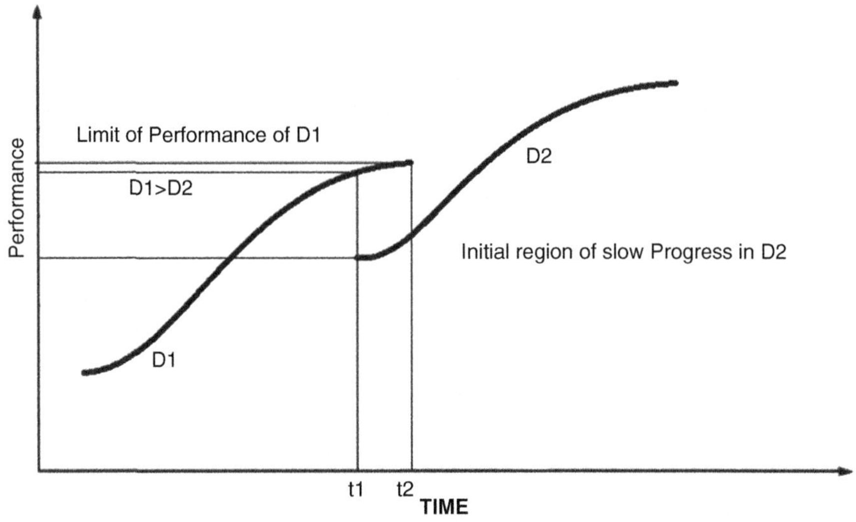

Fig. 1.1 Destination discontinuity model. (Source: Kumar et al. (2018))

according to them, and leave out factors that become obsolete or non-affective. Thus, the way ahead is to move forward with flexible strategic management. Flexible strategic management includes applying strategies such as the continuity-change matrix, HML-VDB analysis, SAP-LAP framework, and different structural modeling techniques like ISM and TISM in decision-making.

1.8 Research Study Area: Himachal Pradesh – The Abode of God

The present research work has been conducted as a case study manner with Himachal Pradesh as a subject area. The rationale for selecting the state is the presence of innumerous tourism opportunities yet the dismal performance of the state in comparison with other states of the country. Not just the tourism sector of the industry is ailing, but the state has likewise recorded the sixth highest unemployment figure (Himachal Pradesh Economic Survey 2018). This condition is alarming because the state is home to various cultures and traditions that might be lost if there prevails a migration to urban cities in search of employment. To have a complete understanding of the study area, this section is divided into segments to provide an overview of the state. The subsections will talk about the economic aspects of the state, and then an environment scanning has been done to identify the major challenges that threaten the state's economy.

1.8.1 Himachal Pradesh at a Glance

Himachal Pradesh is situated in the northern part of India. The total area of Himachal Pradesh is 55,673 km^2 which includes 55,402 km^2 rural area and 271 km^2 urban area. The state has been divided into 12 districts, namely, Bilaspur, Kullu, Kangra, Mandi, Solan, Hamirpur, Lahaul and Spiti, Shimla, Sirmaur, Una, Chamba, and Kinnaur. As per the 2011 census, the majority of the population, i.e., 89.9%, lives in the rural villages. The area-wise division with a population of districts is as follows.

Talking about the economic background of the state, the gross state domestic product (GSDP) of Himachal Pradesh for 2019–2020 (at current prices) is estimated to be Rs 168,972, showing an increase of 12% over the revised estimates for the years 2018–2019. As per the state budget of 2019, the total expenditure for 2019–2020 is estimated to be Rs 44,388 crore, which means a 1.7% increase over the revised estimate of 2018–2019. Total receipts (excluding borrowings) of the states have witnessed a growth of 8% over the last year revised estimates and now stand at an estimate of Rs 35,024 crore. The revenue deficit for the next financial year is targeted at Rs 2342 crore or 1.4% of the gross state domestic product (GSDP). The fiscal deficit is targeted at Rs 7352 crore (4.4% of GSDP). There was an increase in the allocation of funds to education, transport, and water supply, sanitation, housing, and urban development sectors. There has been an increase of 9.2%, as compared to 2016–2017, in the per capita GSDP of the state. The per capita GSDP of Himachal Pradesh in 2017–2018 (at current prices) was Rs 186,778. According to the Fifth Annual Employment-Unemployment Survey (2015–2016), Himachal Pradesh registered an unemployment rate of 10.6% as compared to the all-India level of 5%. This positions the state as sixth highest in terms of the unemployment rate when contrasted with other states (Himachal Pradesh Economic Survey 2018) (Fig. 1.2).

The state government puts special emphasis on rural development; the government has planned to spend Rs 1722 crores in the next fiscal year, providing a 20% boost to the plans. This highlights the focus of the government on rural development. For the above purpose, a new skill development program called "Mukhya Mantri Gram Kaushal Yojna" will be launched which will provide skill training to artisans in traditional arts and help them gain an assured income. For water supply, sanitation, and urban housing development, the government has planned to propose an outlay of Rs 3207 crores, providing a 6% increase in funds allocated. For the development of the transport sector, the government has proposed an outlay of Rs 3561 crores, with an increased outlay of 6% in the next fiscal year. This whole implies that the government has given due weightage to the development of infrastructure in states in urban regions as well (source: Himachal Pradesh Budget Speech 2019–2020, Himachal Pradesh AFS 2019–2020, Himachal Pradesh Demand for Grants 2019–2020; PRS).

Fig. 1.2 Himachal Pradesh: districts and major cities. (Source: http://himachalpradeshtravel.com)

1.8.2 Major Activities Identified Regarding Occupation in Himachal Pradesh

Evidently, the state has been characterized by the prominence of pivotal economic activities, with agriculture standing out as the predominant driver of employment generation. Since around 90% of the number of inhabitants in the state dwells in rural areas, the agriculture sector is the main source of generating employment. The major crops grown in the area include commercial crops and food grains such as maize, pea, potato, ginger, apple, wheat, and others. The main reason for the growth of commercial cropping is the suitability of agriculture as well as the higher prices for the same. According to the government estimates, the food grain production in the state is targeted at 1.67 million metric tons in 2018–2019. The figures for

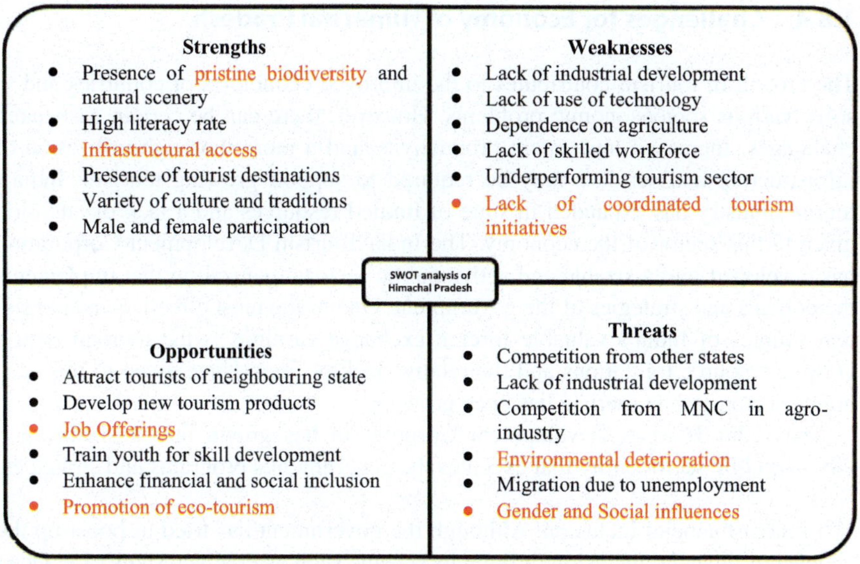

Fig. 1.3 SWOT analysis of Himachal Pradesh. (Source: Author's work)

horticulture production in the state reached 2374.57 thousand metric tons in 2018–2019, as per government estimates. Since the state government announced incentives for the pharmaceutical industries in its industrial policy, this sector has witnessed significant growth. The state has become the fastest-growing region in terms of pharmaceutical industries in the country. During 2018–2019, exports of drug formulations and biologicals from the state stood US$ 533.4 million and have reached US$ 374.6 million between April and September 2019. Moreover, the people are also engaged in industries such as food procurement and processing, light engineering, IT and electronics, cement, tourism and hydropower, and textiles (Himachal Pradesh Economic Survey 2018).

1.8.3 SWOT Analysis of Himachal Pradesh

To completely understand the Himachal Pradesh, a SWOT (acronym for strength, weakness, opportunity, and threat) analysis has been conducted. From the analysis, several challenges have been identified for the state that is either originating from its own limitation (weaknesses) or are a result of outside influence (threats). The analysis presents a mix of the series of steps that the state could undertake based on its own strengths and the opportunities present before it. The emerging challenges are then discussed in the following section (Fig. 1.3).

1.8.4 Challenges for Economy of Himachal Pradesh

The growth of tourism contributes to the improved economies of countries and to solve various socioeconomic problems. However, there can be certain associated challenges. Increased investment productivity and a more thorough utilization of infrastructure and services may be required to support growing tourism. India's tourist industry has expanded in spite of limited resources and a lack of attention given to this sector of the economy. The India Tourism Development Corporation has developed into a sizable and active public sector organization that implements the policies and strategies of the government. One of the most effective and consistent sources of India's valuable foreign exchange earnings is the tourism sector. Despite resource limitations and the relative lack of importance given to this economic sector, Indian tourism has been growing.

The India Tourism Development Corporation has grown into a sizable and vibrant public sector entity that executes the government's programs and strategies.

- Limited financial inclusion: Although the government has tried to boost up the level of financial inclusion through its scheme such as Pradhan Mantri Jan Dhan Yojana, MUDRA, etc., the level of inclusion is far below the level of efforts made into. The very reason for the same is the reluctance of the rural people to engage in financial activities owing to fewer opportunities at all. Rural economies are generally closed in nature. With less of dealing with urban markets, rural market stakeholders find no reason to engage in financial transactions using a proper organized mechanism.
- Limited social inclusion: Himachal Pradesh is a land of tribal and is home to various types of culture, traditions, and customs. These traditions differ not only among districts but within a district one could find variety. The knowledge of the customs, languages, traditions, and food habits is far unknown to outsiders as well as people originating from the state as well. There could be no appropriate way to channelize this information's to the people to educate them about the Himachal Pradesh.
- Dependence on agriculture: Agriculture is the primary activity of people in Himachal Pradesh. A majority of the population living in the state depends on agriculture for their earnings. Since there is dependence on one industry, the economy is under high peril of failure. Since the agriculture sector suffers from its own limitation, failure of one or two seasons can put the livelihood of the number of people in jeopardy.
- Underperformance of the tourism sector: Tourism sector in Himachal Pradesh has not been able to perform up to its latent capacity. This is mainly because of the exhausted marketing mix followed by the tourism policymakers. Common tourism products, failure to attract tourists through marketing campaigns, and inability to tap into opportunities have been identified as problems for the tourism industry. The underperformance is noticeable in the no growth of the state in terms of ranking based on the number of visits by tourists in different states within the country.

- Lack of skilled labor force: Although the literacy rate is particularly high in the state of Himachal Pradesh, not only among males, but concerning females as well, there is a lack of skilled workforce among the population. Service sector industries require skill in dealing with people. It requires updating one's skill to enhance the customer satisfaction. The reason for the lack of skill development is the absence of industries that require a skilled workforce. Dependence on agriculture has promoted the workforce to engage in activities to promote productivity in terms of agriculture and reducing the skilled component of knowledge.
- Unbalanced industrial development: Since a vast area of the state is covered into inhabitable terrain during most of the year, industries can't thrive in such a region. Therefore, few regions have excelled in the industrial development, apart from agriculture, while others are facing difficulty in development. This has caused an imbalance in the development of the districts in the state.
- Migration due to unemployment: Since unemployment is unreasonably high in Himachal Pradesh because of the lack of industrialization and sole reliance on agriculture, there is a threat that people will migrate from the state to other parts of the country in search of better employment opportunities. This will not only lead to substantial economic loss to the state but will also damage the social demographics of the state. It will deplete the social richness of the state and will lead to the loss of tribal culture, ethics, and values for good.

1.9 Rationale of the Study

Within the challenges lie opportunities. The mentioned challenges for the Himachal Pradesh should be addressed so that the state could do wonders in terms of raising the quality of life. These challenges provide the opportunity to explore the realm of the unknown and provide solutions for the same.

- Will help in reducing the burden on agriculture: The study intends to extend rural tourism as an alternate means of earning to people of Himachal Pradesh. This way not only the burden could be reduced on agriculture, but also, certain allied industries could be developed to ease out the burden.
- Promote financial inclusion: The one particular reason for limited financial inclusion was the absence of necessary means to conduct financial transactions because of limited exposure to urban consumers and the market. By inviting urban consumers in your backyard, it will open the way for people in Himachal to interact with them through organized financial sector methods.
- Promote social inclusion: What alternate way could it be better to educate people about your customs, values, traditions, folklore, and food habits than to invite them to participate with you in them. When people will be visiting Himachal Pradesh, they will have an amazing opportunity to have a first-hand experience to know about the same.
- Enhance the performance of the tourism sector by developing products and strategies: Ailing tourism sector requires redesigning. Rural tourism could be a five-

star addition to its product line. Strategies could be designed to promote rural tourism through celebrity endorsement, alter the price mix of the industry, and add destinations as new attractions on the product portfolio.

- Make youth skill equipped: The Government of India has launched schemes such as SANKALP (skill acquisition and knowledge awareness through livelihood promotion) and STRIVE (skill strengthening for industrial value enhancement) to develop skills necessary to conduct activities in the manufacturing sector. When these schemes are combined with the ASPIRE (a scheme for promotion of innovation, entrepreneurship, and agro-products), it creates a weapon to boost up the skill level among youth in rural areas. By offering a service industry that requires skill, youth could be attracted to such schemes.
- Balanced development of the tourism industry across all districts: Each district has destinations that have the potential to be developed as tourism destinations. Since this is an industry that can be developed throughout the state, we infer that it might lead to the holistic and balanced development of rural tourism state.

1.10 Conclusion

The chapter precisely focused on the introduction of the challenges faced by the tourism industry around the world and highlighted the reasons due to which the sector needs the focus of policymakers in the form of the economic contribution made by it. The chapter also tried to present an alternate mode for rural development through the promotion of sustainable rural tourism. The chapter then talked about the challenges faced by the study area, i.e., the state of Himachal Pradesh by conducting an environment scanning and provided justification on how the study will be fruitful in overcoming them. The chapter concludes that rural development schemes have not fully achieved their desired results. Further, till now there is no single industry that could be established all over the country to promote rural development. Because of all these issues, rural development is yet to be achieved completely in India. Although the tourism industry is doing wonders in terms of economic contribution to global GDP and generating employment, simultaneously, the challenges to tourism are hampering the gains from the tourism industry. Overtourism at certain destinations, exhausting marketing mix policies, and poor destination management policies are some of the challenges severely impacting tourism globally. Further, the outbreak of the COVID-19 pandemic will hit the tourism industry really hard in recent years. The chapter presents that practices such as destination discontinuity and practicing flexible management strategies to formulate strategies and policies will help the tourism industry in providing efficient economic gains.

The chapter also presented the demographic and economic aspects of the study area, Himachal Pradesh, a northern state situated in India. The state has several challenges like a high rate of unemployment, migration of people, the presence of unskilled labor, and an ailing tourism sector. The chapter concludes various ways in which this study will help in overcoming shortages of our study area and will generate economic gains to the state.

References

Buckley, R. (2002). Tourism ecolabels. *Annals of tourism research*, 29(1): 183–208.

del Río-Rama, M.D.L.C., C.P. Maldonado-Erazo, J. Álvarez-García, and A. Durán-Sánchez. 2020. Cultural and natural resources in Tourism Island: Bibliometric mapping. *Sustainability* 12: 724.

Government of Himachal Pradesh Economic Survey. 2018. *Economic and Statistics Department, Government of Himachal Pradesh.*

Kumar, S., K. Attri, and S. Shekhar. 2018. Niche tourism: Sustainability and marketing strategies. *Tourism Innovations* 8 (1): 1–9.

Lane, B. (1994). What is rural tourism?. *Journal of sustainable tourism* 2 (1–2), 7–21.

Mir, L.A. 2014. An economic evaluation of Indian tourism industry. *International Journal of Scientific and Research Publications* 4: 1–7.

Zolfani, S.H., M. Sedaghat, R. Maknoon, and E. Zavadskas. 2015. Sustainable tourism: A comprehensive literature review on frameworks and applications. *Economic research-Ekonomska istraživanja* 28: 1–30.

Website Links

https://www.census2011.co.in/census/state/districtlist/himachal+pradesh.html

https://www.nriol.com/india-statistics/himachal-pradesh/districts.asp

https://himachalservices.nic.in/economics/pdf/ImportantStatistics_2014_15.pdf

https://himachalservices.nic.in/economics/ecosurvey/en/agriculture_and_horticulture.html

https://www.ibef.org/states/himachal-pradesh.aspx

http://tourism.gov.in/sites/default/files/Other/India%20Tourism%20Statistics%202018.pdf

http://tourism.gov.in/sites/default/files/Other/Statement%20Containing%20Brief%20Activities%20of%20The%20Ministry%20of%20Tourism%202018.pdf

https://www.thehinducentre.com/the-arena/current-issues/article28680544.ece

https://www.dhanbank.com/pdf/reports/InFocus-December%201,%202010.pdf

Tourism Trends in India and the State of Himachal Pradesh

2

2.1 Tourism Industry in India: An Overview

History of travel industry in India can be followed back to the reported beginning of civilization. With the motto of "Atithi Devo Bhava," India has consistently welcomed the tourists be it from any country. When the Punjab state was created in 1966 and parts of its territory were given to Himachal Pradesh, the history of tourism officially began. The development of tourism in India Corporation was established in October 1966 with the intention of advancing India's tourist industry. Himachal Pradesh Travel Development Corporation was established in 1972 as one of the key forces behind the development of India's tourism industry infrastructure to handle the escalating workload and state-wide development. The culture of India places a high emphasis on the satisfaction of the visitor, even higher than oneself. Since colonial times, the Himalayas have drawn people there because of the resources they contain, which has fostered government use and exploitation of the region. Due to its scenic beauty, recreational opportunities, and the local cottage businesses that have grown over time, coupled with the tourism sector that has allowed the growth and sustenance of the local hill economies, the Himalayan terrain has traditionally been preferred for the tourism industry. The history of tourism in Himachal Pradesh can be found in the historical writings of Hiuen Tsang, a famous Chinese traveler, and later colonial travelers like William Moorcroft, who wrote extensively about the region's social and economic life, art and architecture, natural beauty, and flora and fauna. In addition, being a home of such a vast variety of food, landscapes, languages, cultures, religions, and a history of being an economic and education powerhouse, India has always attracted tourists toward it. Scholars like Hiuen Tsang, Al Biruni, Xuan Zang, and I-Qing visited India during the ancient period to acquire information about cultural beliefs and education policies. Similarly, traders were frequent visitors to India for their commercial interests. There exist a lot of tourism destinations that exist in India which have attracted the

S. Kumar, *Sustainable Rural Tourism in Himalayan Foothills*,
https://doi.org/10.1007/978-3-031-40098-8_2

interests of tourists not only domestically as well as across the globe. These destinations are spread across all the states and union territories of India. The table below lists down the specialty of each of the states concerning tourism and also mentions about renowned tourism destinations in each state and union territories (Table 2.1).

2.1.1 Trends in Tourist Arrival in India

Keeping a systematic record of the number of foreign as well as domestic tourists is of significance for any country. Not just it helps in calculating the footfall of tourists, it also helps in designing product mix, but it has necessary economic as well as social implications. Tables 2.2 and 2.4 present the number of tourists both international and domestic visiting India over the years.

Table 2.2 highlights the positive trends in the international tourist's visits over the years. This upward trend is the result of pro-international tourist policies followed by the Government of India. The scheme of visa on arrival for tourists from around more than 50 countries has made it easier for tourists to visit India. As per the data, a total of 10.56 million foreign tourists and 6.87 million NRIs visited India for various purposes. These figures registered a growth of 5.18% and 1.4% p.a. respectively as compared to previous years. The arrival of foreign tourists is always welcomed by any country. This is on the grounds that such an arrival has economic as well as social impact for any country. When a foreign tourist visits a country, they bring with them foreign exchange earnings. Similarly, it additionally helps in the exchange of technology and products between tourists and natives. This aids in raising the social standards and also the lifestyle of individuals living in a region.

Figure 2.1 demonstrates the patterns in foreign tourist arrival in India over the years 2001–2018. A trend line has been plotted and a linear equation is derived for the trend line. The figure has several important observations. To begin with, there has not been any significant outflow in the number of tourist's arrival in the country over the years. Second, the number of visits failed to beat the trends during the period 2009–2015. This might be because of the global depression and terrorist attacks over the years. However, the figure likewise features that the industry is getting back on tracks and the figures for the last 2 years have beaten the patterns by a decent edge. This indicates the good health of the tourism industry (Suneel et al. 2018).

Although foreign tourists are important for economic and social well-being, no tourism industry can survive without the support of domestic tourists. The presence of domestic tourists is essential not only for developing infrastructure, but they, by visiting places throughout the year, help in maintaining the infrastructure as well. The livelihood earned by the locals depends much upon the domestic tourists rather than foreign tourists. Table 2.2 indicates the trends in the domestic tourist visits in all the states and union territories (UTs) over the years. As per the data, it indicates that approximately 1854.93 million people traveled to all the states during the years 2018–2019, registering an annual growth of 11.91% p.a. compared to the previous year. It recommends the blossoming condition of economy as well as industry.

Table 2.1 Famous tourism destinations across different states and union territories in India

Sr. No	Name of state/ UT	Specialty	Major tourism destinations
1.	Andaman and Nicobar Island	Scenery, sightseeing, adventure tourism	Elephant beach boat ride, scuba diving at Havelock Island, Jolly Buoy island trip, Chidyatapu trek trip, Cellular jail visit in Port Blair, etc.
2.	Andhra Pradesh	Religious tourism, sightseeing, pilgrimage, movie tourism	Tirupati temple, Araku valley, Srisailam Mallikarjuna, Papikondalu, Vijayawada, Chittoor, Talakona waterfalls, Ramoji film city, etc.
3.	Arunachal Pradesh	Sightseeing, leisure	Tawang, Bomdila, Ziro, Itanagar, Namdapha national park, Roing, Tezu, Parshuram Kund, etc.
4.	Assam	Sightseeing, adventure, and religious tourism	Kaziranga National Park, Kamakhya temple, Umananda temple, Pobitora Wildlife Sanctuary, Rang Ghar, Majuli island, etc.
5.	Bihar	Religious sites, food, handicrafts, wellness sites, and sightseeing	Bodhgaya, Mahabodhi Temple, Vishnupad Temple, Mangla Gauri Shrine, Bodhi Tree, Chinese Temple and Monastery, Nalanda, Relic Stupa, Takht Sri Patna Sahib, Hieun Tsang Memorial Hall, Kundalpur, etc.
6.	Chandigarh	Sightseeing	Rock garden, Zakir Hussain garden, Sukhna lake, government museum and art gallery, etc.
7.	Chhattisgarh	Sightseeing, leisure, wellness tourism	Chitrakote waterfalls, Bhoramdeo Temple, Bhilai steel plant, Jagdalpur, Maikal Hills, Dantewada, Kanger Valley National Park, etc.
8.	Dadra and Nagar Haveli	Leisure and adventure tourism	Vanganga lake, Khanvel, Swaminarayan Temple, Dudhani Lake, Tribal Museum, Hirwa Van garden, etc.
9.	Daman and Diu	Religious, cultural, and leisure tourism	Naida caves, Ghoghla Beach, Diu Fort, Gangeshwar Temple, Nagoa Beach, St. Paul's Church
10.	Delhi	Spiritual, leisure, religious, cultural, dark, medical, MICE, adventure, sports tourism	Red Fort, Qutub Minar, Akshardham temple, tombs of emperors, Hauz Khas village, Rajiv Chowk (CP), multispecialty hospitals, trade fairs and exhibitions, India Gate, Rajghat, etc.
11.	Goa	Leisure, religious, cultural, wedding, and adventure tourism	Calangute beach, Baga beach, Agonda beach, various churches and historical monuments, Basilica of Bom Jesus, Palolem beach, etc.
12.	Gujarat	MICE, religious, cultural, spiritual, sports, leisure, and adventure tourism	Kutch, Sabarmati Ashram, Lakhota museum, Bhavnath Mahadev, Somnath temple, Gir national park, Dwarka, Junagarh, Saputara Hills, Motera stadium, etc.
13.	Haryana	Food, leisure, religious, eco, rural, cultural, and dark tourism	Badkhal Lake, Morni Hills, Murthal, Sohna Lake, Damdama Lake, Kurukshetra, Panchkula, Gurgaon Cybercity, Ambience mall, Sultanpur Bird Sanctuary, Panipat, Karnal, etc.

(continued)

Table 2.1 (continued)

Sr. No	Name of state/ UT	Specialty	Major tourism destinations
14.	Himachal Pradesh	Leisure, wellness, spiritual, religious, sports, movie, wedding, and adventure tourism	Shimla, Manali, Mcleodganj-Dharamshala, HPCA cricket stadium, Dalhousie, Spiti valley, Kasol, Mandi, Shikari Devi temple, Kullu valley, etc.
15.	Jharkhand	Leisure, cultural, eco, rural tourism	Ranchi, Hazaribagh Wildlife Sanctuary, Betla National Park, Parasnath, Bokaro steel plant, Jamshedpur, Deoghar, Canary Hill, Netarhat, etc.
16.	J&K	Leisure, sightseeing, and religious tourism	Sri Nagar, Gulmarg, Tulip Garden, Vaishno Devi temple, Sonamarg, Gulmarg, Amarnath, Patnitop, Pahalgam, Dal Lake, Leh and Ladakh, etc.
17.	Karnataka	Ecotourism, religious tourism, MICE, and sightseeing	Keshava Temple, Coorg Hills, Gokarna, Chikmagalur, Bangalore, Udupi, Mysore Palace, Hampi, Gol Gumbaz, etc.
18.	Kerala	Sightseeing, leisure, spiritual, wellness, adventure, cultural, and religious tourism	Alleppey backwater, churches in Kerala, Munnar, Kumarakom, Trivandrum beaches, Kerala Folklore Theatre and Museum, Athirappilly, etc.
19.	Lakshadweep	Sightseeing and adventure tourism	Minicoy Island, Kavaratti Island, Amindivi Island, Agatti Island, Lighthouse, Kalpeni, etc.
20.	Madhya Pradesh	Spiritual, religious, cultural, wellness, rural, eco, and dark tourism	Khajuraho erotic temples, Gwalior forts and caves, Buddhist monuments, Taj ul Masajid, Sanchi stupa, Orchha Fort, Bandhavgarh, etc.
21.	Maharashtra	Reality, dark, religious, cultural, sightseeing, leisure, wellness, MICE	Gateway of India, Dharavi slums, Mumba Devi Temple, Taj Hotel, Juhu beach, Bhandardara, Panhala Hills, Ajanta Caves, Jain temple, Siddhivinayak temple, Pune—a hill station, Nasik, Khandala, Lonavla, etc.
22.	Manipur	Religious, sightseeing, and adventure tourism	Imphal, Thoubal, Sri Sri Govindajee Temple, Loktak Lake, Tamenglong, Sirohi National Park, Manipur State Museum, Chandel, Moirang, etc.
23.	Meghalaya	Religious, leisure, sightseeing, and adventure tourism	Butterfly Museum, Umiam Lake, Mawsmai Caves, Cherrapunjee, Elephant Falls, Mawsynram, Tura, etc.
24.	Mizoram	Sightseeing and adventure tourism	Serchhip, Durtlang Hills, Champhai, Lunglei, Dampa Wildlife Sanctuary, Saiha, etc.
25.	Nagaland	Religious, sightseeing, and adventure tourism	Dimapur, Phek, Mon, Naga Heritage Village, Kohima War Cemetery, Mokokchung, Kisama Heritage Village, Kohima Museum
26.	Odisha	Spiritual, religious, cultural, wellness, rural, and ecotourism	Rayagada, Cuttack, Baripada, Lingaraja Temple, Hanuman Vatika, Sun temple, Mahanadi Barrage, Puri Beach, Jagannath Temple, etc.

(continued)

Table 2.1 (continued)

Sr. No	Name of state/ UT	Specialty	Major tourism destinations
27.	Puducherry	Sightseeing and adventure tourism	Aurobindo Ashram, Seaside Promenade, Church of the Sacred Heart of Jesus, Paradise Beach, Pondicherry Botanical Garden
28.	Punjab	Food, leisure, religious, eco, agro, rural, cultural, pilgrimage, spiritual, and dark tourism	Attari border, Jallianwala Bagh, Patiala, Ludhiana, Pathankot, Akal Takht, Rangla Punjab Haveli, Golden temple, Bhatinda Zoological Park, Rural Olympics at Kila Raipur, etc.
29.	Rajasthan	Food, leisure, religious, eco, agro, rural, cultural, pilgrimage, spiritual, and dark tourism	Jaisalmer, forts and palaces in Udaipur, Hawa Mahal Jaipur, Mount Abu, Jodhpur, Thar desert, Jal Mahal, Umaid Palace, Bikaner, Ranthambore, Chittorgarh, Pushkar, Bharatpur, etc.
30.	Sikkim	Sightseeing and adventure tourism	Zuluk, North Sikkim, Gangtok, Tsomgo Lake, Ranka Monastery, Do Drul Chorten Stupa, etc.
31.	Tamil Nadu	Food, leisure, religious, eco, agro, rural, cultural, pilgrimage, spiritual, dark, medical, MICE, and spiritual tourism	Rameshwaram, Yercaud, Vivekanand memorial, Bragadeeshwara Temple, Mahabalipuram monuments, Ooty Botanical gardens Kanyakumari, Chennai, Mahabalipuram, Madurai, Kanchipuram, etc.
32.	Telangana	Halal, religious, MICE, leisure tourism	Hyderabad, Papikondalu, Warangal, Nagarjuna Sagar, Nizamabad, Secunderabad, Chilkur Balaji temple, Gayatri waterfalls, etc.
33.	Tripura	Spiritual, sightseeing, and adventure tourism	Unakoti, Kailashahar, Jagannath Temple, Iskcon Radha-Govinda Mandir, Neermahal Palace, Bison national park, Rudrasagar lake, etc.
34.	Uttar Pradesh	Food, cultural, pilgrimage, religious, adventure, MICE, leisure, sightseeing, spiritual, wedding tourism	Taj Mahal, Vrindavan temple, Imambara in Lucknow, Varanasi, Prayagraj, Mathura temples, Sarnath temples, Vindhyachal, Chitrakoot, Ayodhya, Dudhwa National Park, Jhansi, Kanpur, Noida city, Kushinagar, Loni, Firozabad, Bareilly, Fatehpur Sikri, Nawabganj bird sanctuary, Dewa Sharif, etc.
35.	Uttarakhand	Leisure, wellness, spiritual, religious, sports, movie, wedding, and adventure tourism	Dehradun, Rishikesh, Auli, Nainital the lake district, Gun Hill, Mussoorie, Jim Corbett National Park, Haridwar, Kedarnath Dham, Badrinath Temple, etc.
36.	West Bengal	Religious, spiritual, cultural, leisure, food, and ecotourism	Darjeeling, Kolkata, Sundarban, Siliguri, Digha Sea Beach, Murshidabad Historical City, Kurseong, Haldia, Santiniketan, etc.

Source: Author's compilation

Table 2.2 Foreign tourist arrival in India over the years

Year	Foreign tourist arrivals (in millions)	Percent change	Domestic tourist arrival (in millions)	Percent change
2001	2.54	−4.02%	236.47	7.43%
2002	2.38	−6.30%	269.6	14.01%
2003	2.73	14.71%	309.04	14.63%
2004	3.46	26.74%	366.27	18.52%
2005	3.92	13.29%	392.04	7.04%
2006	4.45	13.52%	462.44	17.96%
2007	5.08	14.16%	526.7	13.90%
2008	5.28	3.94%	563.03	6.90%
2009	5.17	−2.08%	668.8	18.79%
2010	5.78	11.80%	747.7	11.80%
2011	6.31	9.17%	864.53	15.63%
2012	6.58	4.28%	1045.05	20.88%
2013	6.97	5.93%	1142.53	9.33%
2014	7.68	10.19%	1282.8	12.28%
2015	8.03	4.56%	1431.97	11.63%
2016	8.8	9.59%	1615.39	12.81%
2017	10.04	14.09%	1657.55	2.61%
2018	10.56	5.18%	1854.93	11.91%

Source: Ministry of Tourism report, Government of India report 2018

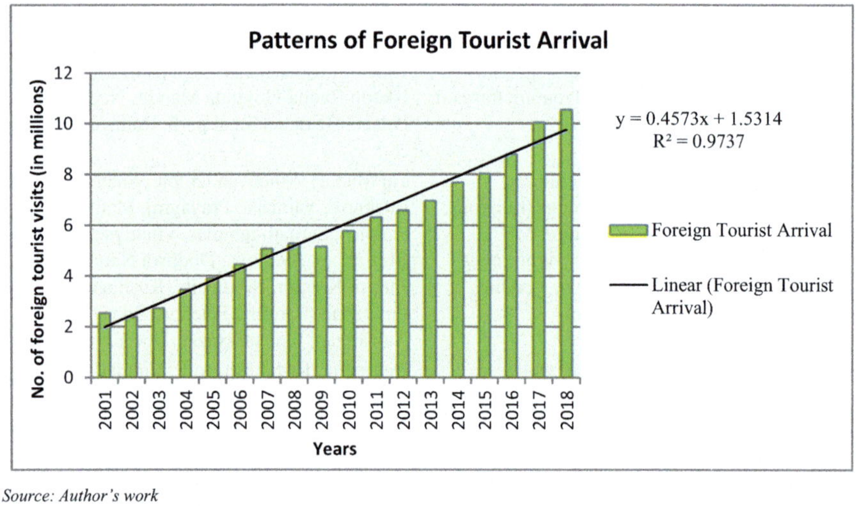

Fig. 2.1 Patterns of foreign tourist arrival in India. (Source: Author's work)

When people are happy and are prospering economically, they tend to enjoy their life. Touring and traveling are how people tend to enjoy in their life.

Figure 2.2 indicates the patterns of domestic tourist visits in India. The graph points to two interesting observations. First, it is noticeable that the number of

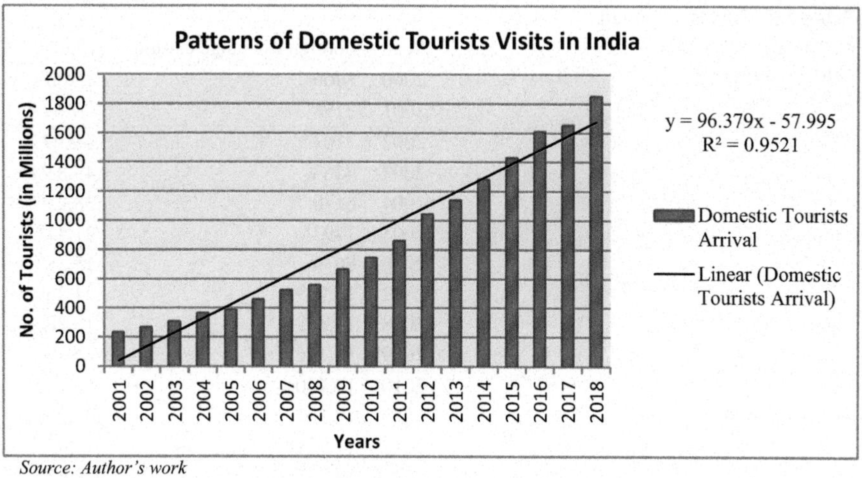

Source: Author's work

Fig. 2.2 Patterns of domestic tourist visits in India. (Source: Author's work)

domestic tourist visits falls below the linear trend line during the period of 2006–2013. During this period, the country was rocked by the terrorist attacks and was suffering from the shocks of the global recession, although the shock was negligible and did not cause much depression during the time. Second, since 2014, the domestic tourist visits have always beaten the trend estimates for the year. This shows that the domestic tourism industry is booming across the country. On the off chance that we incorporate the tourism industry because of the Kumbha Mela and other recently organized tourism events in the country, at that point the numbers would be definitely more than what it is currently. These are the indications of healthy tourism industry. It is interesting to take note of that the government did not mention the income earned from domestic tourists over the years in its annual tourism statistics. It is a very significant statistic as the contribution of domestic tourists cannot be overlooked, and without recognizing the contribution of domestic tourists, not much strategy can be developed to ensure and maintain the flow of tourists.

2.1.2 Trends in Foreign Exchange Earnings in India

When foreign tourists visit India, they contribute to the foreign exchange reserves of the country. These contributions are very significant as they help in setting up of infrastructure at destinations and also fill up the reserves for the other activities of a country. The following table lists down the foreign exchange earnings of India over the previous years.

Table 2.3 presents the absolute measure of foreign exchange earnings from the arrival of foreign tourists over years 2000–2018. Foreign exchange reserves are of very much importance for a nation. It not only impacts the value of the exchange

Table 2.3 Trends in foreign exchange earnings from tourism in India

Year	Foreign exchange earnings	Percent change
2000	3460	15.00%
2001	3198	−7.57%
2002	3103	−2.97%
2003	4463	43.83%
2004	6170	38.25%
2005	7493	21.44%
2006	8634	15.23%
2007	10,729	24.26%
2008	11,832	10.28%
2009	11,136	−5.88%
2010	14,490	30.12%
2011	17,707	22.20%
2012	17,971	1.49%
2013	18,397	2.37%
2014	19,700	7.08%
2015	21,013	6.66%
2016	22,923	9.09%
2017	27,310	19.14%
2018	28,585	4.67%

Source: Ministry of Tourism, Government of India, report 2018

rate, but it is also required for meeting international transactions. Tourism has played a vital role in foreign exchange earnings in India. In the year 2018, India earned USD 28,585 million from the foreign tourist's arrival.

Figure 2.3 indicates the patterns for foreign exchange earnings by India from foreign tourist arrival over the period 2000–2018. The figure provides three observations. First, the actuals have tended to just match up the estimates over the years. Second, there was a wide gap between the estimates and the actual figure during 2009. This was mostly because of global depression. Third, the estimates were again higher than actuals in the years 2015 and 2016. Yet, the industry is again on the right track as the actual foreign exchange earnings over the years have beaten the estimates by the margin during 2017 and 2018. This clearly shows that foreign tourists are arriving and spending a lot of money as well. This is essential for the industry and economy to bloom.

2.1.3 Relationship Between Foreign Tourists Arrival and Foreign Exchange Earnings

It is often argued that the foreign tourist arrival (FTA) and the foreign exchange earnings (FEE) from foreign tourists are not perfectly correlated. As discussed by Suneel et al. (2018), despite the fact that there exists a relationship between foreign

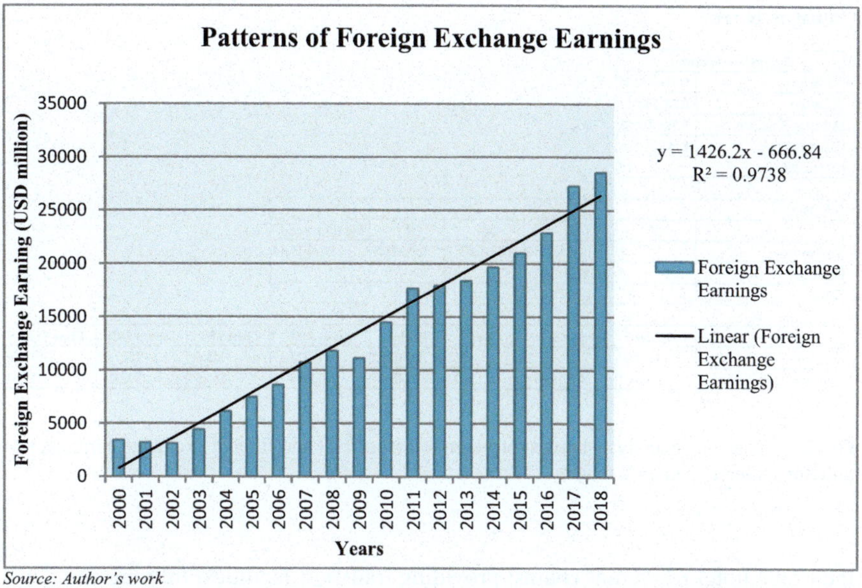

Fig. 2.3 Patterns of foreign exchange earnings from tourism in India. (Source: Author's work)

tourist's arrival and foreign exchange earnings, they are not perfectly positively correlated. Several factors impact the foreign exchange earnings from FTA. So, in this section, we attempted to analyze the relationship between FTA and FEE through a single linear regression analysis.

Figure 2.4 proposes that the change in FTA explains a 98.91% change in the value of FEE over the years. This confirms the observation made by Suneel et al. (2018) that change in FTA is not solely responsible for the change in FEE. Other factors such as the length of tourist stay, foreign exchange rate, spending power of the foreign tourist due to the global economic environment, and attractiveness of destination also impact the tourists spending power directly or indirectly.

2.2 Role of Institutional Support for Development of Tourism Industry in India

Individual elements by themselves are insufficient to fully account for the performance and adaptation of expatriates. An industry cannot work without institutions providing to it necessary support. These institutions serve as a backbone for the industry by providing it with policies, regulations, and strategies for its development. An institution could be government or nongovernment; it could be global, national, or state-based; or it could be developed by major private players in the industry. These institutions perform a wide plethora of services that include running

SUMMARY OUTPUT

Regression Statistics	
Multiple R	0.994568576
R Square	0.989166652
Adjusted R Square	0.988489568
Standard Error	856.0147535
Observations	18

ANOVA

	df	SS	MS	F	Significance F
Regression	1	1070506372	1070506372	1460.92109	3.7437E-17
Residual	16	11724180.13	732761.2581		
Total	17	1082230552			

	Coefficients	Standard Error	t Stat	P-value	Lower 95%	Upper 95%	Lower 95.0%	Upper 95.0%
Intercept	-4688.07	532.77	-8.80	1.58E-07	-5817.49	-3558.66	-5817.49	-3558.66
Foreign Tourist Arrivals	3207.63	83.92	38.22	3.74E-17	3029.73	3385.54	3029.73	3385.54

Source: Author's work

Fig. 2.4 Regression analysis between foreign tourist arrival and foreign exchange earnings from tourism. (Source: Author's work)

their own hotel restaurant chains; providing transport facilities; training and development of tour guides, stakeholders, etc.; and conducting market research and statistics gathering. In order to provide better insights into expatriate adjustment and performance, we have included organizational elements (reward and support) in addition to individual factors (risk-taking and resilience). Be it any type, institutions working for promotion of tourism in India serve several following objectives:

- To promote the development of the tourism industry to assist in the economic development of the country/state
- To enhance the role of the tourism industry in the sociocultural development of the region
- To increase international trade through the participation of tourists in trade fairs, exhibitions, and other avenues
- To develop strategies to increase the inflow of tourists both international and domestic
- To protect the interests of tourists and the tourist industry
- To ensure the existence of a tourist market so that a continuous supply of tourists can be maintained that is necessary for sustaining the infrastructure developed in a region
- To promote research, training, and development in the field of tourism
- To symbolize tourism as a means of promoting goodwill, harmony, and peace among people in turbulent times

There are four levels at which institutions support the development of tourism in India. Various agencies are working on the desired objectives under each category. Although their scale of operations might be different, the essence of their work can be summarized in the objectives discussed above.

- International agencies: The organizations working at the international level intend to develop tourism with a view of creating understanding peace, prosperity, and per human rights and fundamental freedoms for all without any discrimination. These organizations work to increase the inflow of tourists for the socioeconomic development, strengthen sociocultural ties between countries, and create a barrier-less market for tourists. Agencies such as World Tourism Organization (WTO), World Tourism and Travel Council (WTTC), Pacific Asia Travel Association (PATA), and International Air Transport Association (IATA) are international agencies working for the development of tourism industry.
- National-level agencies: The organizations working at the national level aim to create a market for the development of the tourism industry at a country level. These agencies create strategies, lay out plans with visions and missions, and conduct country-based research and database management. These institutions layout strategies for training and development of stakeholders involved in different schemes, master out the budget for future activities to be conducted in the year, and other similar activities. The Ministry of Tourism, India Tourism Development Corporation (ITDC), Hotel Corporation of India Limited (HCIL), and India Tourism Transport Undertaking Limited (ITTUL) are the national-level agencies working for tourism promotion in India.
- State-level agencies: These establishments work for integrated development of the tourism industry in India by cooperating with the adjoining states and union territories. These agencies execute the plans and strategies created by the national-level agencies and create a smooth barrier-less market for the state. These agencies set up state plans and draw state budgets for the industry. In India, states and union territories have Directorates of Tourism and Tourism Development Corporations set up in 18 states and union territories of India. Himachal Pradesh Tourism Development Council is a state-based organization for the development of its tourism industry.
- Nongovernment agencies: These organizations are basically nongovernment organizations (NGOs) working for the promotion of the tourism industry in India. Several organizations are created as an implied or express agreement among the private key players in the industry that promotes self-regulations on the industry. These regulations help in creating a smooth working environment and to keep any check from the state, national, or international agencies.

2.3 Initiatives of Government Schemes for Tourism Industry

The Government of India or state/union territory governments, through ministry of tourism or state tourism development boards, time and again formulate schemes for the betterment of the tourism industry. These schemes may be a part of their tourism policy or may be standalone schemes. These schemes might be related to a particular segment like niche tourism development, training and development, schemes for financial assistance, or for the holistic development of the industry. Some of the key schemes for the tourism industry at the national level are listed below:

- Scheme for rewarding travelers visiting 15 destinations in a year.
- Swadesh Darshan scheme
- PRASAD—the National Mission for Pilgrimage Rejuvenation and Spiritual Augmentation Drive
- "Hunar Se Rozgar Tak" scheme
- E-Visa on arrival for nationals of 50+ countries
- Adarsh Smarak scheme
- Swatch Bharat Scheme for cleanliness of tourism destination
- Scheme for support to public-private partnership in infrastructure development
- Scheme for organizing fair and festival and tourism-related events (DPPH)
- Scheme of capacity building for service providers (institutes)
- Marketing development assistance (MDA)

2.4 Ranking of States and UTs in Terms of Tourist Arrival in India

Every year, the government in its annual statistics presents a ranking of states and union territories based on domestic and foreign tourist arrival. This ranking is very important because it helps in identifying the underperforming states and the states or UTs that are performing exceptionally well in terms of tourism. This ranking can also be used as a parameter to decide whether the expenditure incurred in planning for tourism has proved to be worthwhile or not. In addition, it helps the states and UTs to identify their potential competitors in terms of attracting tourists. Indeed, performing states can be studied by underperforming states to boost up their industry as well.

Table 2.4 highlights that Tamil Nadu, Uttar Pradesh, Karnataka, Andhra Pradesh, Maharashtra, and Telangana are performing admirably in the event of pulling in domestic tourism. Domestic tourism has thrived in these states mainly due to the presence of archaeological wonders and religious sites. Tourist-friendly policies and the presence of basic infrastructure have also significantly contributed to tourist footfall. Presently, in the event if one talks about the ranking as per foreign tourist arrival, then Maharashtra bags the top position followed by Tamil Nadu, Uttar Pradesh, and Delhi. This is a result of the presence of international airports that have connectivity to almost every country in the world. In addition, since Mumbai is the commercial hub of India, MICE tourism is also practiced here. This ranking also highlights the fall of states such as Jammu and Kashmir, Uttarakhand, and Assam in ranking over the last decade. Similarly, Himachal Pradesh has remained stagnant in its ranking. This highlights that tourism policy practiced in the state has failed to attract tourists of the neighboring state.

Table 2.4 Ranking of states and union territories in India in terms of tourist arrival

		Ranking of states and UTs over the year											
		2006		2008		2011		2013		2015		2017	
Sr. no	Name of state/UT	D	F	D	F	D	F	D	F	D	F	D	F
1.	Andaman and Nicobar	30	24	31	23	31	25	30	25	32	27	30	27
2.	Andhra Pradesh	1	7	1	7	2	13	3	14	3	16	4	15
3.	Arunachal Pradesh	32	32	30	30	30	28	32	27	31	31	31	29
4.	Assam	17	23	18	22	20	24	21	24	22	26	22	26
5.	Bihar	11	16	11	11	11	7	11	8	13	8	14	9
6.	Chandigarh	21	21	21	20	22	22	24	22	25	25	25	25
7.	Chhattisgarh	26	31	27	32	26	31	10	30	16	29	20	30
8.	Dadra and Nagar Haveli	23	30	25	26	28	32	28	33	29	34	29	34
9.	Daman and Diu	24	25	26	27	24	29	25	29	26	30	28	31
10.	Delhi	18	1	19	1	13	3	14	3	15	4	15	4
11.	Goa	19	10	20	10	21	11	22	10	23	10	21	10
12.	Gujarat	9	14	9	14	10	14	8	16	9	14	9	13
13.	Haryana	15	17	17	16	19	16	20	13	21	13	23	17
14.	***Himachal Pradesh***	***12***	***11***	***12***	***9***	***14***	***10***	***16***	***11***	***17***	***12***	***16***	***12***
15.	Jharkhand	20	26	16	25	15	19	17	19	11	17	13	18
16.	J&K	13	19	13	18	16	18	13	20	20	22	18	22
17.	Karnataka	4	8	10	13	4	9	4	9	4	9	3	11
18.	Kerala	14	9	14	8	17	8	18	7	18	7	17	8
19.	Lakshadweep	34	29	35	31	35	35	35	35	36	35	36	36
20.	Madhya Pradesh	10	13	5	12	6	12	6	12	7	11	8	14
21.	Maharashtra	6	2	6	2	5	1	5	1	5	2	5	1
22.	Manipur	31	35	32	35	32	34	31	32	33	32	33	33
23.	Meghalaya	25	27	23	28	25	27	26	28	27	28	27	28
24.	Mizoram	33	33	33	34	33	33	33	34	34	36	34	35
25.	Nagaland	35	34	34	33	34	30	34	31	35	33	35	32
26.	Odisha	16	20	15	19	18	20	19	18	19	21	19	21
27.	Puducherry	22	18	24	24	23	21	23	21	24	19	24	20
28.	Punjab	27	12	22	17	12	15	12	15	14	15	11	7
29.	Rajasthan	5	5	4	5	7	5	7	5	10	6	10	5
30.	Sikkim	28	22	28	21	27	23	27	23	28	23	26	24
31.	Tamil Nadu	3	4	3	3	3	2	1	2	1	1	1	2
32.	Telangana	–	–	–	–	–	–	–	–	6	18	6	16
33.	Tripura	29	28	29	29	29	26	29	26	30	24	32	23
34.	Uttar Pradesh	2	3	2	4	1	4	2	4	2	3	2	3
35.	Uttarakhand	7	15	7	15	8	17	15	17	12	20	12	19
36.	West Bengal	8	6	8	6	9	6	9	6	8	5	7	6

Source: Ministry of Tourism reports, 2006–2018
Note: *D* Domestic Tourists, *F* Foreign Tourists

2.5 Tourism Industry in Himachal Pradesh: An Overview

The travel industry in Himachal Pradesh is famous since the time of the colonial rule. British used the hills of the state as a relief measure during the summers in plains. Shimla was established for the same purpose and was nicknamed "summer capital of India." During British rule, tourism in Himachal Pradesh was restricted to a few destinations and was primarily used for leisure purposes. But once India gained independence, since then, the tourism industry in Himachal Pradesh has not looked back. Due to its presence in the Himalayan belt, Himachal Pradesh offers a wide range of topology and weather to its tourists. In addition, there is a wide range of flora and fauna present in the state that acts as a tourist attraction. Further, it is home to the residence of Dalai Lama, which attracts tourists from all around the globe. The presence of extant diversity in topology, culture, weather, and flora and fauna has enabled the tourism industry to thrive in Himachal Pradesh. One of the most lucrative sectors for the state is swiftly emerging as the tourist industry. The tourism industry in Himachal Pradesh makes a significant contribution to the gross domestic product. It is estimated that 6% of the total GDP is contributed by the tourism and hospitality sector. The tourism industry also provides employment to a large scale of people directly and indirectly. As per Kumar et al. (2018), three districts in Himachal Pradesh account for about 50% of the tourist's inflow. These districts are Kullu, Shimla, and Kangra. The following table lists the prominent tourist destinations in these districts. These three districts are regarded as a major tourist hub in the city. They have destinations that can cater to almost every typology of tourists from adventure to religious to leisure. The following table lists down the major tourism destinations in these three districts of Himachal Pradesh (Table 2.5).

Since these districts are famous for tourism, subsequently, these always gather the attention of tourists. Yet, certain destinations can act as alternate for the famous tourist destinations of the state. These unexplored or underexplored destinations are present throughout the state and in every district. The following table lists down the underexplored or unexplored tourist destinations present in the different districts of the state and also presents the major sites for which these could go about as a substitute (Table 2.6).

2.5.1 Tourist Arrival Patterns in Himachal Pradesh

Himachal Pradesh has always been a popular choice among the foreign as well as domestic tourists. Due to the presence of hill stations, people tend to visit the state to get relief from high-soaring temperatures during the summers in plain regions. People who tend to enjoy snowfall visit the state during the winters to celebrate their Christmas and welcome New Year simultaneously. Apart from these, religious tourists are welcomed by the state throughout the year. The following table lists down the inflow of tourists, both foreign and domestic, in Himachal Pradesh during the period 2008–2018.

Table 2.5 Prominent tourist destinations in Shimla, Kullu, and Kangra districts of Himachal Pradesh

District	Adventure tourism sites	Religious tourism sites	Leisure tourism sites
Shimla	Ice skating at circular road	Bhimakali temple	Ridge
			Mall road
			Kufri
			Theog
			Scandal point
		Jakhu temple	Green valley
		Kalibari temple	
		Christ church	
	Rafting in Tattapani	Tara Devi temple	
	Skiing in Kufri and Narkanda		
	Fishing in Rohru and Seema		
Kullu	River rafting	Hadimba Devi Temple	Kasol
		Bijli Mahadev Temple	Great Himalayan National Park
		Raghunath Temple	
	Paragliding	Manikaran	Tirthan valley
	Trek to Malana village	Vashist	
	Bhrigu lake trek		
	Kheerganga trek		
	Chandrakhani pass		
Kangra	Trekking in Minkiani pass, Indrahar pass	Jwalamukhi Temple	Kangra art museum
			Dal lake
			Bhagsu falls
			Kangra Fort
		Ambika Devi temple	Tea gardens
	Water sports in pong dam	Bajreshwari Devi temple	
	Angling in Dehra and pong dam reservoir	Kaleshwar Mahadev temple	
	Aero sports in Bir, Billing and Dhelu	Chamunda Devi Mandir	

Source: Annual Reports of Himachal Pradesh Tourism Development Corporation (2018)

Table 2.7 suggests that the flow of tourists has often been dwindling in the state both for domestic and foreign tourists. If the state has witnessed an increase in a year, then it is followed by a decrease in the upcoming years. This reduces the benefits that a state could generate from tourist inflow. Talking about foreign tourists, the state welcomes 356,568 foreign tourists during the year 2018, which is approx. 25% less as compared to the previous year. This underperformance could be attributed to the crisis of water shortage which occurred during the year. Similarly, the state welcomed 16,093,935 domestic tourists, registering a downfall of approximately 16% when compared with a year ago. This shows the ailing situation of the tourism industry in Himachal Pradesh (Fig. 2.5).

Table 2.6 Few unexplored and underexplored tourist destinations in Himachal Pradesh

District	Under/Unexplored Tourism Destinations	Could be alternate to
Chamba	1. Churah valley 2. Pangi valley	Green Valley, Tirthan Valley
	3. Manimahesh temple	Religious Tourism sites
	4. Khajjiar	Sightseeing and Religious Tourism
	5. Kalatop wildlife sanctuary	National parks in Kullu
Kinnaur	1. Bhaba pass 2. Sangla valley 3. Lippa 4. Chitkul	Valleys and hills of Shimla.
	5. Reckong Peo	Shopping in Mall road.
	6. Kalpa	Buddhist tourism sites in other districts
	7. Moorang	Kangra fort, Sightseeing.
Mandi	Religious Tourism destinations	
	1. Kamru Nag	Religious tourism sites in other districts
	2. Shikari Devi 3. Sundernagar Sheetla Mata Temple	Naina Devi temple.
	Sight Seeing destinations	
	4. Karsog valley	Valleys in Kullu
	5. Barot	Destinations of Kangra
	6. Rewalsar Lake	Lakes of Shimla
	Adventure Tourism destinations	
	7. Janjheli	Similar destinations in Shimla, Kangra
	8. Pandoh Dam	Rafting in Kullu, Kangra
	Niche Tourism destinations	
	9. Devidarh	Wellness Tourism in other districts
	10. Prasar Lake	Shooting movies in Shimla and Kangra
	Tourism Circuit Development	
	11. Chindi	Chindi can act as a religious tourism site as well as a sightseeing destination for tourists visiting Kangra. A complete circuit for community-based tourism could be developed here.
Lahaul & Spiti	Lahaul Valley-	
	1. Triloknath temple	Temples in other districts
	2. Keylong	Sightseeing destinations in Shimla
	Spiti Valley-	
	1. Key Monastery	Developed as a hub of

(continued)

Tab. 2.6 (continued)

	2. Tabo Caves and valley 3. Chandra Taal Lake 4. Kibber	Buddhist tourism and wellness tourism in the state. Adventure tourism can be promoted once the connectivity issue is resolved.
Bilaspur	1. Bhakra Nangal Dam 2. Gobind sagar Lake 3. Water sports and fishing 4. Koldam Dam	Adventure tourism sites in Kangra, Kullu and Shimla
	5. Sri Naina Devi ji temple	Religious Tourism sites
Hamirpur	1. HP Technical University 2. National Institute of Technology 3. Institute of Hotel Management	MICE & education tourism in Shimla and Kangra.
	4. Sujanpur Tira	Historical tours in Shimla
	5. Baba Balak Nath temple	Religious tourism sites
Una	1. Mata Chintpurni Devi temple	Pilgrimage tourism in HP
	2. Kaleshwar temple	Religious Tourism sites
	3. Thaneek Pura 4. Kila Baba Bedi Ji	Cultural, leisure tourism destinations of Shimla, Kangra
Solan	1. Shoolni Mata temple	Religious tourism sites
	2. Mohan Park	Leisure tourism destinations
	3. Karol Tibba	Sightseeing in Shimla and Kullu
	4. Motilal Nehru state Library	Education tours in HP
Sirmaur	1. Churdhar	Religious tourism in HP
	2. Renuka Lake	Sightseeing destination
	3. Haripur	Heritage tourism in HP
	4. Habban Valley	Food tourism (Peach) development center

Source: Annual Reports of Himachal Pradesh Tourism Development Corporation (2018)

Table 2.7 Inflow of tourist in Himachal Pradesh over the years

Year	Domestic Tourists	% change in Domestic Tourists	Foreign Tourists	% change in Foreign Tourists	Total Tourists	% change in total tourists
2008	9,372,697	–	376,736	–	9,749,433	–
2009	11,036,572	17.75%	400,583	6.33%	11,437,155	17.31%
2010	12,811,986	16.09%	453,616	13.24%	13,265,602	15.99%
2011	14,604,888	13.99%	484,518	6.81%	15,089,406	13.75%
2012	15,646,048	7.13%	500,284	3.25%	16,146,332	7.00%
2013	14,715,586	−5.95%	414,249	−17.20%	15,129,835	−6.30%
2014	15,924,701	8.22%	389,699	−5.93%	16,314,400	7.83%
2015	17,125,045	7.54%	406,108	4.21%	17,531,153	7.46%
2016	17,997,750	5.10%	452,770	11.49%	18,450,520	5.24%
2017	19,130,541	6.29%	470,992	4.02%	19,601,533	6.24%
2018	16,093,935	−15.87%	356,568	−24.29%	16,450,503	−16.08%

Source: Reports, Department of Tourism & civil aviation, Himachal Pradesh

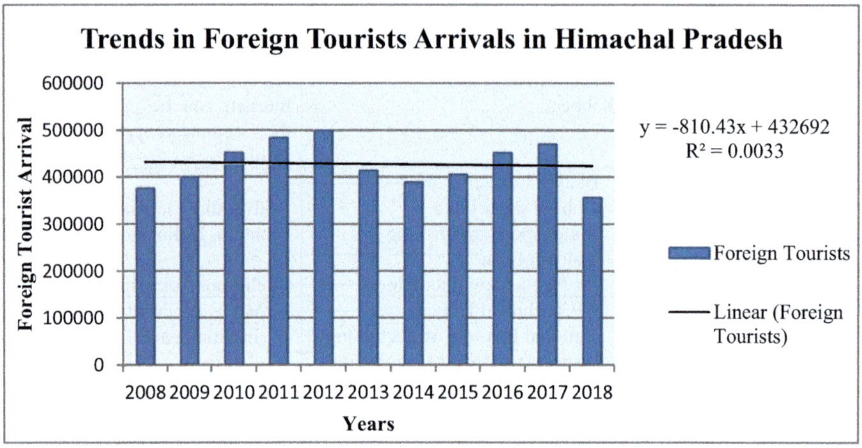

Source: Authors' work

Fig. 2.5 Trends in foreign tourist arrival in Himachal Pradesh. (Source: Authors' work)

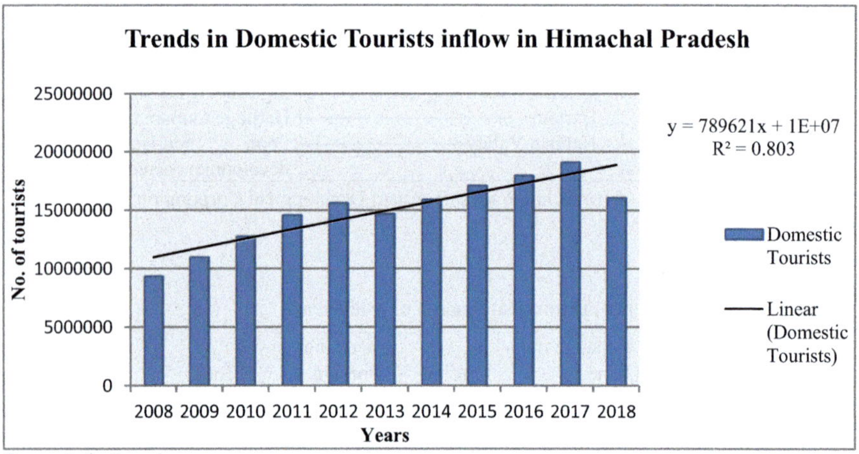

Source: Author's work

Fig. 2.6 Trends in domestic tourist arrival in Himachal Pradesh. (Source: Author's work)

The above graph highlights the trends in foreign tourist arrival in the state over the period 2008–2018. A linear trend line has been drawn from the figures. The graph presents the underperformance of the tourism industry in the state. The actual inflow of tourists failed to beat the estimates in the most of the years under study. In addition, whenever the state progressed, it was followed by a sudden decrease in the number of tourists in subsequent years, vanishing out any gains that aroused out in earlier years (Fig. 2.6).

The above graph indicates the trends of domestic tourist inflow in Himachal Pradesh throughout 2008–2018. The trends for the same are not encouraging as the

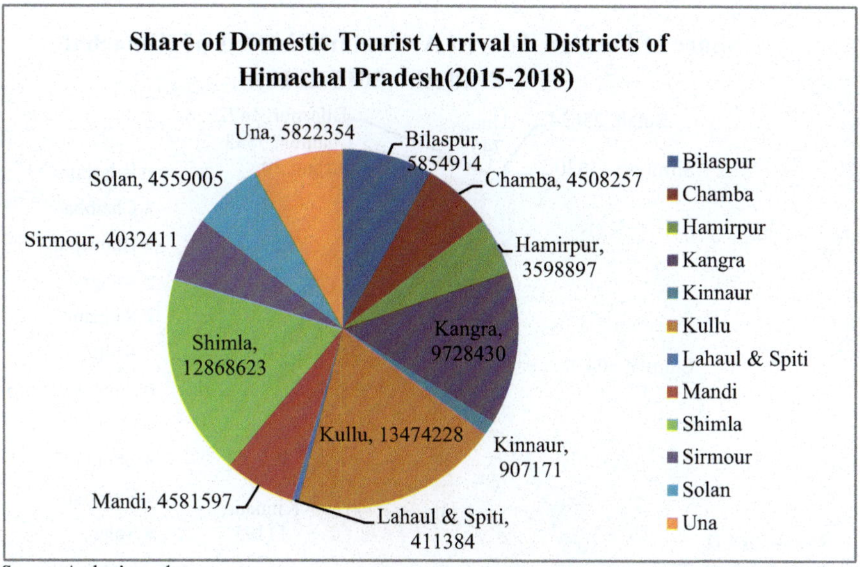

Source: Author's work

Fig. 2.7 Share of domestic tourist arrival in various districts of Himachal Pradesh. (Source: Author's work)

graph suggests that the industry has often barely managed to just touch the estimates. And in a few years, the actuals were less than estimates. This shows the problematic situation in the industry as if an industry is unable to attract domestic tourists than how can it think of attracting foreign tourists. Additionally, domestic tourists are responsible to provide funds to maintain infrastructure at any place. Without domestic tourists coming, infrastructure maintenance will be troublesome for the policymakers.

2.5.2 District-Wise Dynamism of Tourist Arrival in Himachal Pradesh

The following section manages introducing a district-wise analysis of the flow of foreign and domestic tourists in Himachal Pradesh. The table below lists the flow of tourists in every district of Himachal Pradesh during the period of 2015–2018 (Fig. 2.7).

The graph above shows that the flow of domestic tourists over the years has been distributed evenly among a few states. The graph also suggests that Kullu, Shimla, and Kangra take lead in the share of domestic tourist arrival in the state. However, the share of Kinnaur and Lahaul and Spiti is minimal (Fig. 2.8).

The above graph suggests that the inflow of foreign tourists over the period 2015–2018 has been profoundly skewed in a few states only. The inflow is restricted

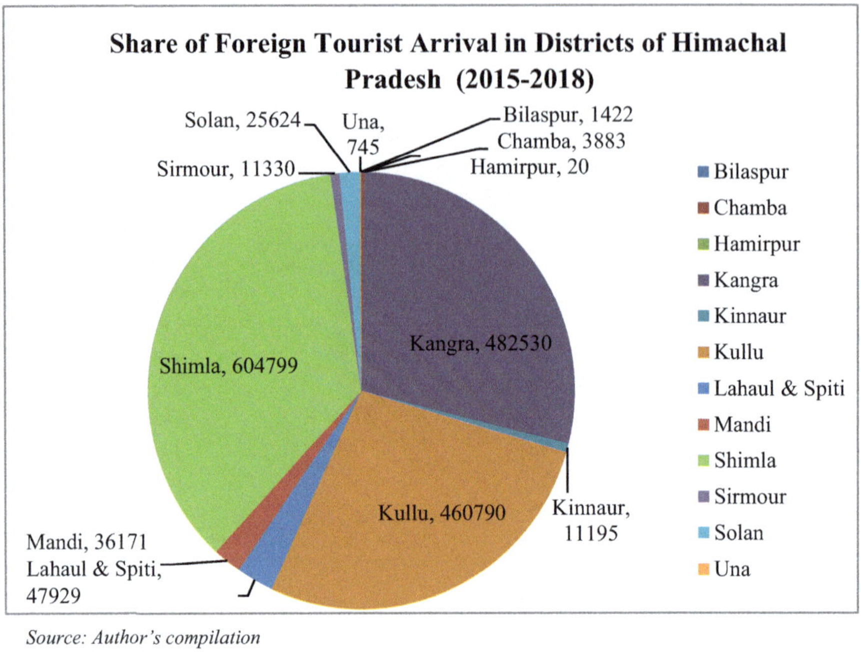

Fig. 2.8 Share of foreign tourist arrival in Districts of Himachal Pradesh. (Source: Author's compilation)

to Shimla, Kullu, and Kangra, and other districts have a negligible contribution to the inflow. This shows that there is an imbalance in the development of tourist destinations for foreign tourists in the state.

2.5.3 Emerging Registered Stakeholders in Himachal Pradesh Tourism Industry

In the event that tourists comprise of the demand side of the tourism industry, there is a supply side for the same. The supply side consists of locals living in the state and other stakeholders. These stakeholders are vital for the survival of the industry as they act as key players in the industry. The policies developed should keep in mind the interest of stakeholders as well. The other stakeholders in the tourism industry include hoteliers, homestay owners, restaurant owners and staff, travel agencies, photographers, tourist guides, and adventurers. Each of these stakeholders is crucial for enhancing the satisfaction level of tourists. In addition, a state needs to maintain a record for the number of stakeholders operating in its industry. Table 2.8 is a district-wise list of registered stakeholders in the tourism industry in Himachal Pradesh.

Table 2.9 suggests that an aggregate of 3349 registered hotels operate in Himachal Pradesh. Out of these, the majority of hotels are located in Kullu, Kangra, and

Table 2.8 District-wise tourist arrival in Himachal Pradesh

Sr. No	Districts	2015		2016		2017		2018	
		Domestic	Foreign	Domestic	Foreign	Domestic	Foreign	Domestic	Foreign
1.	Bilaspur	1,368,807	350	1,431,831	440	1,616,925	446	1,437,351	186
2.	Chamba	1,122,894	1197	1,162,267	823	1,180,949	828	1,042,147	1035
3.	Hamirpur	855,263	4	906,914	12	980,611	2	856,109	2
4.	Kangra	2,396,970	112,843	2,528,837	123,895	2,684,948	138,341	2,117,675	107,451
5.	Kinnaur	117,216	2695	97,864	2863	463,771	2609	228,320	3028
6.	Kullu	3,314,463	109,468	3,515,169	122,064	3,732,044	133,057	2,912,552	96,201
7.	Lahaul and Spiti	86,591	4612	100,759	15,278	104,645	14,275	119,389	13,764
8.	Mandi	1,086,231	10,478	1,152,614	11,181	1,231,968	10,072	1,110,784	4440
9.	Shimla	3,261,152	154,155	3,416,629	165,476	3,318,829	162,168	2,872,013	123,000
10.	Sirmour	1,016,060	3377	1,059,618	3558	992,352	2522	964,381	1873
11.	Solan	1,072,486	6692	1,117,837	6973	1,225,105	6454	1,143,577	5505
12.	Una	1,426,912	237	1,507,411	207	1,598,394	218	1,289,637	83
	Sub Total	17,125,045	406,108	17,997,750	452,770	19,130,541	470,992	16,093,935	356,568
	Total	17,531,153		18,450,520		19,601,533		16,450,503	

Source: Annual Reports, Department of Tourism and Civil Aviation, Himachal Pradesh

Table 2.9 Number of registered stakeholders of the tourism industry in Himachal Pradesh

Sr. No.	District	No. of hotels	Homestay units	No. of restaurants	Travel agencies	Photographers	Tourist guides	Adventurers
1.	Bilaspur	84	14	11	28	1	29	2
2.	Chamba	225	106	35	51	17	106	5
3.	Hamirpur	60	7	32	6	0	2	0
4.	Kangra	672	268	109	231	3	128	97
5.	Kinnaur	73	66	9	19	0	12	0
6.	Kullu	953	327	72	1374	783	386	73
7.	Lahaul and Spiti	88	258	30	15	0	1	0
8.	Mandi	198	100	42	115	4	75	38
9.	Shimla	493	334	125	978	91	451	6
10.	Sirmour	111	42	69	7	0	55	0
11.	Solan	286	125	152	81	0	69	0
12.	Una	106	9	17	7	0	0	0
	Total	**3349**	**1656**	**703**	**2912**	**899**	**1314**	**221**

Source: www.himachaltourism.gov.in

Shimla. This might be since the demand for hotels is high in these districts. The number of hotels is low in Hamirpur, Bilaspur, and Kinnaur. Likewise, there are a total of 1656 homestay units in Himachal Pradesh. The number of homestays in the state has expanded drastically due to the emphasis of the government over setting up homestay units at tourist destinations. However, Hamirpur, Sirmaur, and Una have very few homestay units. If we talk about restaurants, then, the state has 703 registered restaurants serving a variety of cuisines including traditionally Himachali food. A great deal of travel agencies operate throughout the state. The majority of travel agencies are found in Kullu and Shimla. Through these agencies, the tour and travel packages for all the districts could be availed easily. The state has about 899 registered photographers, of which 783 are found exclusively in the district of Kullu. However, the advent of smart mobile phones has made it very difficult for photographers to sustain. The state has around 1314 registered tourist guides who impart knowledge to the tourists about local destinations, themes, folklore, customs, traditions, and food. In addition, the state is home to 221 adventurers whose objective is to build the satisfaction level of adventure tourists.

2.6 Himachal Pradesh Tourism Policy 2019: An Assessment

Himachal Pradesh tourism policy 2019 states its vision as "Positioning Himachal Pradesh as a leading global sustainable tourism destination for inclusive economic growth." This implies that the objective of the state government is to promote tourism sustainably and also it wants that the economically backward section should also enjoy the benefits of the industry. For this purpose, the policy highlights several missions:

- Protection of the state's natural and cultural heritage
- Improved quality of life and better employment opportunities
- Enhanced tourist experience
- Innovation through private sector participation

The policy divides the objectives into several segments. In the first place, it talks about product diversification in the form of developing destinations on a theme-based model. Second, it aims to protect tourist destinations by introducing the concept of sustainability in tourism. Third, it ensures that the policies are pro-stakeholders and benefit the natives of the state. Fourth, it talks about building up the capacity of the human resources available for the industry by providing them with training and development opportunities. Fifth, it promotes inclusiveness by developing products suitable to need of every section of society. And last, it encourages the participation of the private sector in the development of the tourism industry.

To accomplish the above-discussed objectives, the state follows several strategies. These strategies are developed with different time horizons of up to 10 years. To start with, destinations that are underexplored and have tourism potential in terms of niche tourism segments like eco, agro, wellness, MICE, lake, snow, heritage and culture, pilgrimage, film, and adventure will be developed on a 5-yearly basis. Then a promotional strategy will be followed to inform tourists about these circuits. The issue here with the strategy lies in its essence of marketing itself to tourists. Mere advertising on tourist websites will not help in attracting tourists. Without any advertising campaign, the reach of the tourist circuits will be limited to a few. This might be a challenge in terms of the sustainability of circuits in the long run. The next strategy talks about improving the accessibility of destinations through the development of roadways and airways. The policy also talks about the development of smart toilets, Wi-Fi zones, water ATMs, smart parking, etc. to provide tourists world-class facilities. This strategy is a very much welcomed step as tourist's particularly foreign class requires high-class infrastructure at destinations. Notwithstanding, the challenge will be to ensure the maintenance of the infrastructure as there will be a need to maintain a balance between uses of these services by domestic tourists and foreign tourists. The third strategy to be adopted by the state includes the use of innovative marketing promotions schemes such as celebrity endorsement, geographical indication tagging, endorsement through film tourism, and so forth. This is a very much appreciated step as, without an effective advertising campaign, the industry has failed to attract tourists. However, there has been no work done in this respect until now. The next strategy involves undertaking carrying capacity-based tourism destinations development. This strategy will be crucial to promote sustainable tourism, but the policymakers must keep the social, economic, and technological aspects of carrying capacity in mind as well. The next strategy talks of incorporating green practices like waste management, resource conservations, environment audits, etc. This is also a welcomed step but guidelines need to be formulated for the same, and strict adherence to it must be ensured. The next strategy involves providing community support for the development of tourism and ensuring that the flow of tourists is maintained throughout the year. The flow of tourists must be maintained continuously so that the burden to maintain infrastructure does not fall on the locals and natives. The next strategy is to develop the required human resources to enhance tourist satisfaction. Certain niche tourism being developed will require specialized skilled personnel. By providing the youth with the required skills, the problem of unemployment can also be tackled, and migration to cities could be stopped. For this, the government can make use of schemes such as STRIVE, SANKALP, and ASPIRE to train and develop skillful human resources for the tourism industry. The next strategy involves making tourism safe for all through the use of ICT (information and communication technologies) in tourism destination management. The only hurdle in front of this strategy is the lack of infrastructure present to permit the utilization of ICTs at destinations. The last strategy adopted for achieving the objective is to attract private players in the development of tourism through FDI. However, there will be a problem of incentives for private players that could lure them into this industry. Since the industry is not performing to the expectations, private players might be hesitant in entering the industry. Although by including tourism expenditure in the list of corporate social responsibility activities, private players could be incentivized to work for the industry.

2.7 Tourism Development: The Role of Himachal Pradesh Tourism Development Corporation (HPTDC)

The Himachal Pradesh Tourism Development Corporation (HPTDC) was formed in September 1972 as an undertaking of the Government of Himachal Pradesh under the company act 1956. The Himachal Pradesh Tourism Development Corporation has acted as a catalyst, trendsetter, and a prime mover for the promotion of domestic and foreign tourism in Himachal Pradesh. It has a wide network of accommodation that comprises of hotels, restaurants, and cafeterias serving varied cuisine, including Himachali delicacies. HPTDC also operates a fleet of Volvo and deluxe buses to facilitate the tourists visiting within and outside the state. HPTDC also operates its own tours and travel packages such as its famous four trekking circuit programs to attract in tourists.

HPTDC was set up to carry out activities for the development of tourism within the state. These activities included establishing, developing, promoting, executing, operating, and carrying on projects, schemes, and other activities including running and maintenance of tourist vehicles. Further, the establishment of interstate and intrastate tourist information centers was also one of the objectives of HPTDC. It was likewise mindful to advertise the travel industry of the state by publishing periodicals to be sold nationally and internationally. The corporation was also to be responsible for promoting the culture, fairs, temples, tourist attractions, and cuisines of the state. The Himachal Pradesh Tourism Development Council performs the following functions:

- Promotion of state tourism regionally, nationally, and globally
- Establishing hotels, restaurants, and resorts at major and also at minor tourist destinations

- Improving the transportation facilities of the railways, roadways, and airways across the state
- Maintaining various heritage sites and historical sites of the Himachal Pradesh
- Developing various tourism sites or destinations of the Himachal Pradesh
- Organizing various cultural events, fairs, exhibitions, and trade fairs with the help of the local community
- Establishing various theme parks and eco-parks in different parts of the state
- Organizing various exhibitions and meetings related to the promotion of tourism

2.8 Assessment of Opportunities and Challenges of Tourism Industry in Himachal Pradesh

To have a better understanding of the present and future trends for the tourism industry in Himachal Pradesh, environmental scanning has been done using SWOT analysis. This analysis helps in a detailed and in-depth assessment of hurdles, challenges, and opportunities for the tourism industry of Himachal Pradesh. SWOT is an acronym for strength, weakness, opportunity, and threat analysis. Under this, the factors impacting the tourism industry of the state will be categorized into one of the categories mentioned, and afterward their impact will be discussed along with probable mitigation strategies (Fig. 2.9).

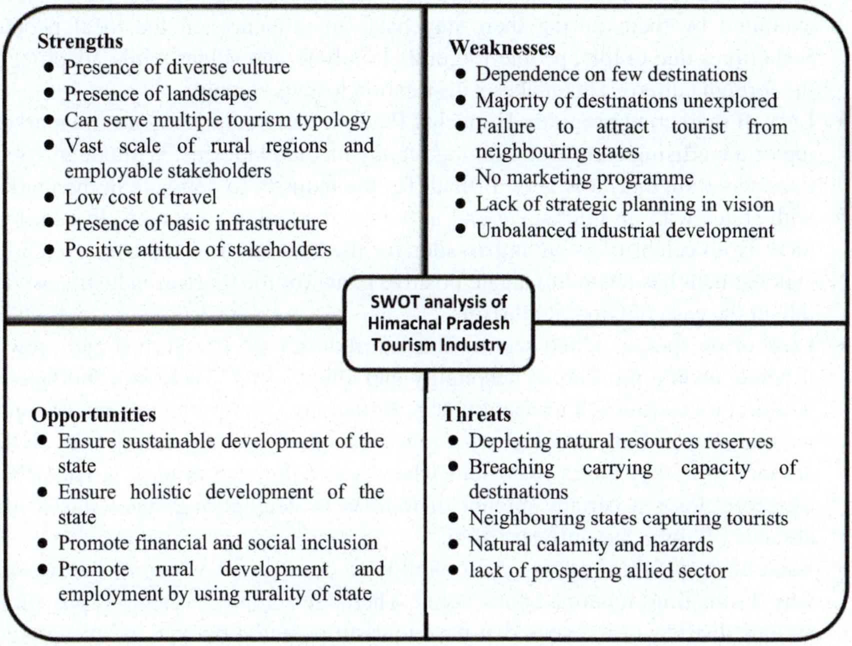

Source: Author's work

Fig. 2.9 SWOT analysis of Himachal Pradesh tourism industry. (Source: Author's work)

2.9 Prospects and Challenges of Tourism Industry in Himachal Pradesh

The tourism industry in Himachal Pradesh is not performing as per its true potential. This is visible in the fall of tourist visits last year and fall in earnings of the state from tourism. The reasons for the underperformance are the challenges in the form of weaknesses or threats to the industry identified by the SWOT analysis. The major challenges for the tourism industry in Himachal Pradesh are as follows:

- Depleting natural resources: There is a decrease in the number of tourists during the year 2018 as compared to the previous year. One of the main reasons is the decreasing level of nonrenewable natural resources in the state. Last year, there was widespread protest by the locals in Shimla asking the tourists to refrain from visiting the district because of a lack of drinking and usable water. Such incidents are not good for the image of the industry.
- Breaching carrying capacity: Every destination has carrying capacity, i.e., there is a limit to which it can accommodate people due to limit on natural and man-made infrastructure. Certain destinations in the district of Shimla, Kangra, and Kullu have frequently witnessed a breach of carrying capacity during peak season. This has resulted in the destruction of tourist destinations. Thus, the economic gains from over-tourism are reducing due to the damage it is causing to the environment and ecology. Further, carrying capacity has a social phenomenon as well. The visits of tourists, particularly foreign tourists, and the behavior exhibited by them during their stay have an influence on the local people. Sometimes due to this, people forget their culture and values while adapting to the foreign culture. This makes a destination lose its serenity.
- Lack of marketing program: Himachal Pradesh tourism industry has no marketing or advertising campaign running on any media platforms. Without an effective marketing mix, it is very difficult for the industry to compete in the market with states with an interactive and attractive marketing campaign. In addition, there is no celebrity as an ambassador for the state tourism industry. Celebrity endorsement has tended to create positive gains for the tourism industry, as visible in the case of Gujarat tourism.
- Lack of prosperous allied sector: Tourism industry for its survival and growth depends upon a prosperous hospitality and allied sector. There is a shortage of world-class facilities in terms of hotels, restaurants, transportation, and communication sector. Certain category of tourists looks for luxury during stay. In the absence of it, they prefer to go somewhere where they can avail of it. Himachal, therefore, loses a certain amount of tourists to neighboring states due to the absence of high-class infrastructure.
- Issue of unexplored regions: Accessibility to destinations is one of the reasons why destinations remained unexplored. There are certain destinations present in various districts of the state that have tourism potential but yet are unexplored. Mandi, before a year or two, was one such state with many underexplored tourism destinations. However, it caught the attention of the government, and the

industry has soon witnessed some progress over there. But, still, various destinations remain in the hope that 1 day the policymakers will soon take notice of it.
- Lack of strategic planning in vision: Policies and strategies drawn for the tourism industry are rigid in nature. So plans drawn become obsolete when there is any change in the environment impacting the industry. Because of the lack of strategic management in the policymaking, the plans are not effective in solving the concerns.
- Lack of natural hazard and calamity: There is a risk that a disaster like the Himalayan tsunami can appear and destroy the tourism industry in Himachal Pradesh. Incidents such as landslide and accidents occur frequently in hilly areas of the state. If these incidents occur frequently or at a large scale, it can cause damage to the industry.

The tourism industry ought to be viewed as sustainable in the event that it added in the accompanying ways:

1. Providing a long-term prosperity with less incurred environmental costs, rather than the momentary advantages
2. Ensure that locals gain generous benefit from it
3. Help to reinforce and enhance the neighborhood economy
4. Reduction in exacerbating implications at community and national levels
5. Assuming these things can be accomplished, then, at that point, there could be a true accord between rural tourism and economy.

2.10 Implications of the Study for Tourism Industry in Himachal Pradesh

Any study needs to have implications in the form of benefits to society. Without a doubt, tourism contributes significantly to Himachal Pradesh's economic growth and offers a development alternative. But because of the government's lack of focus and the absence of fundamental infrastructure development, the state's tourist industry is constantly suffering, making it difficult for Himachal Pradesh to compete in the local and worldwide tourism markets. The Himachal Pradesh economy depends heavily on tourism, so every effort must be made to keep it alive and grow it further. One of the principle objectives of the study is to boost the development of the tourism industry in Himachal Pradesh. The previous chapter has suggested two significant conclusions. First, rural development is the engine for the economic development of the country and rural tourism, and second, the tourism industry can be a tool for rural development. Therefore, through this study, the tourism industry of Himachal Pradesh is focused, and strategies are to be framed for its development. Through this, the other objectives of the study will be achieved simultaneously. The study will help in overcoming the challenges in several ways:

- Challenge of depleting natural resources: One of the main advantages of Himachal Pradesh, identified in the first chapter, was the presence of a vast rural region in the state. This study is based on the promotion of rural tourism for the development of rural development. Since rural regions encompass immense resources lying unused, this study will help in increasing the resource base of the state by utilizing the rural resources and reducing the usage of natural assets in urban areas.

- Concern for breaching carrying capacity: Certain destinations have exhausted their carrying capacity in the district of Kullu, Kangra, and Shimla. By shifting the burden of tourists from such destinations, this study will help in reducing the burden of destinations by following the process known as destination discontinuity. It will be beneficial for the industry as the exhausted destinations can get time to rejuvenate themselves and serve tourists more effectively and efficiently.
- Issue of under-explored regions: To release the pressure from existing destinations, new destinations with tourism potentials will have to be identified. This implies that new or underexplored destinations with rural tourism potential will be identified for the purpose of the development of rural tourism. This will help in the development of regions that are yet to be explored and developed properly.
- Lack of marketing program: Rural tourism and its sub-products will add to the marketing mix of the Himachal Pradesh tourism industry. A marketing program will be formulated with or without celebrity endorsement as the need may be to attract and inform tourists about the products offered by the state in tourism industry. With a bunch of new niche tourism products, the industry will be in a good position to attract tourists.
- Lack of prosperous allied sector: For developing rural tourism, basic infrastructure will be developed at various destinations. Since the burden of mass tourism will ease out at present destinations, they will be in a position to develop them exclusively for a rich class segment, and for that they will have world-class facilities. Similarly, the development of rural tourism will infuse opportunities for the allied sector to start their ventures and grow. In this manner, the allied sector will be able to develop at places where it is hesitant to work due to a lack of opportunities.
- Lack of strategic planning in vision: The study undertakes flexible strategies such as interpretive structure modeling, continuity-change matrix analysis, and others to formulate the strategic crystals for the development of rural tourism in the state (Kumar et al., 2019). These strategies will be flexible in nature and will incorporate any environmental changes in the future. The models developed can be generalized for other niche tourism segments as well.
- Opportunity for sustainable development of state: Sustainable rural tourism requires developing ways to carry out tourism by causing minimal damage to the environment and resources. Thus, the study will help in the sustainable development of the state. This will ensure the endurance of the industry in the long run and will be beneficial for the local people of the state, economically, socially, as well as ecologically.
- Opportunity for the holistic development of state: Rural tourism can be developed in every district of the state, even at places where industrial development is

not possible because of the lack of resources. This is because the rural area is present in abundance in every district of Himachal Pradesh. So, this development of the service sector will help in restoring the balance of development in the state, and the fruits of development could be enjoyed in every district of the state.

- Promote rural development through financial and social inclusion: Use of innovation, technology, and development of communication infrastructure will assist the financial sector to flourish in the state. Thus, the economy could move toward a higher level of financial inclusion due to the availability of opportunities to consume financial resources at ease. Similarly, by inviting tourists to their backyards, locals can inform and educate them about their customs, values, culture, and traditions and furthermore could gain from them about their lifestyle as well. This will enhance the level of social inclusion in the state.

Creating a novel, sustainable travel industry by uniting unique highlights accordingly provides a fascinating experience for the traveler out of previously existing sights and services.

2.11 Conclusion

The chapter makes certain significant conclusion that suggests the rationale for undertaking the study. According to the HPTDC report 2018, there has been found a significant fall in the number of tourists visiting the state. The contribution of the tourism sector to state GDP is quite significant. The travel industry in Himachal Pradesh is being recognized as one of the most important sectors of the economy and a major engine of growth in the economy. But, the expenditure made by foreign tourists is not found to be generating much revenue for the development of the economy. In addition, there was inconsistency found in database management of tourists as no record of earnings from domestic tourists was found in statistics made available by the state through the concerned department. Although the state is endowed with geographical and cultural diversity, clean environment, snow-clad mountains, gushing rivers, historic monuments, and the friendly and hospitable people, tourism in the state is unable to flourish even after the schemes and initiatives such as "Aaj purani rahon se." Whatever development has been undertaken has largely been limited to the urban destinations. These urban destinations have either exhausted or are on verge of exhausting and are unable to serve the tourist effectively. In addition, there is a high lopsidedness in the development of industry within the districts. These drawbacks have caused the ranking of Himachal Pradesh in terms of the number of tourist arrival both domestic and foreign, to fall when compared to other states. The following table shows the ranking of the state over 10 years (Table 2.10).

On the other side, rural destinations having huge tourism potential have largely remained underexplored. Accordingly, there is a need for formulating strategies that would help in tapping the vast potential of rural tourism available with the state.

Table 2.10 Ranking of Himachal Pradesh over the years in terms of tourist arrival

Year	2006	2007	2008	2010	2011	2012	2013	2014	2015	2016	2017	2018
Rank (DTA)	12	12	10	13	14	15	16	17	17	16	16	16
Rank (FTA)	11	11	09	09	10	10	11	12	12	12	12	12

Source: Ministry of Tourism, H.P. Report 2005, 2006, 2007, 2008, 2010, 2011, 2012, 2013, 2014, 2015, 2016, 2017, 2018
Note: *DTA* domestic tourist arrival, *FTA* foreign tourist arrival

What is required is to ensure that the development of rural tourism in Himachal Pradesh occurs sustainably. State is well known for orchards of apples, so while developing rural tourism, it should be ensured that while creating the necessary infrastructure, there is no harm caused to the orchards. Thus, development should be done while acknowledging and preserving the natural beauty of the destination. Another advantage that rural tourism development would have on the state would be an economic benefit that it would provide in rural areas of the state. The rural population would be provided with an alternate source of earning. This would diminish their reliance on the turbulent and volatile agriculture sector. In addition, this would generate high-scale employment in the rural areas, which in recent times has been a challenge for the government. This would put a check on the widespread migration of people from rural areas to urban areas in search of jobs. It would ensure savings and investment habits among the rural people and would thereby necessitate the use of financial instruments by them. This is in line with the government's objective of financial inclusion. Not only financially, rural tourism would ensure social inclusion as well. By inviting tourists to their rural homeland, people can display what truly Himachal Pradesh is all about. From the customs, traditions, history, folklore to authentic Himachali food, there is a lot the world wants to not only know but experience it first-hand. Such experience would be fruitful for the local population in Himachal too.

References

Himachal Pradesh Tourism Development Corporation. 2018. *HPTDC*. Retrieved 2018, from HPTDC Web Site: http://hptdc.in/

Kumar, S., K. Attri, and S. Shekhar. 2018. Niche tourism: sustainability and marketing strategies. *Tourism Innovations* 8 (1): 1–9.

Kumar, Suneel, Navneet Guleria Shekhar, and N. Guleria. 2019. Understanding dynamics of niche tourism consumption through interpretive structure modeling. *Saaransh RKG Journal of Management* 11: 40–48.

Suneel, K., A. Shekhar, and A. Kamlesh. 2018. The economic contribution of tourism industry in India: an appraisal. *International Journal of Applied Hospitality and Tourism Research*: 1–12.

Website Links

http://www.transindiatravels.com/bihar/tourist-places-to-visit-in-bihar
https://traveltriangle.com/blog/places-to-visit-in-andaman/
https://traveltriangle.com/blog/tourist-places-in-assam/
https://www.holidify.com/state/andhra-pradesh/top-destinations-places-to-visit.html
https://www.tourmyindia.com/blog/top-20-must-see-destinations-arunachal-pradesh/
https://www.tourplan2india.com/state-wise-popular-tourist-places-in-india/

Research Framework 3

3.1 Introduction

The research methodology comprises two terms: research and methodology. Research is defined as the systematic and scientific search for information on a particular subject matter. In common terms research is known as the search for new knowledge and further innovations in the existing work on focused topics, while the methodology is described as clear thought on methods and procedure used by researcher to analyze critical information on topics or subject matter to achieve the defined objectives. In other words, research is information collection on crucial areas, and methodology is a track of process and method used to achieve the research objective. Research methodology is an organized way of touching on solving research problem. It helps in drawing primary assumptions and making logical conclusion on predefined research issue. In order to gain insights of research area, an appropriate research methodology technique is necessary to engage in research study (Table 3.1).

The research design of the study is of applied research. Applied research is the practical application of basic research. It includes the assessment of solving a practical problem by using empirical methodologies (de Mello and Pedroso 2018). An exploratory design is conducted about a research problem when there are few or no earlier studies to refer to. There are limited studies available on the conceptualization of rural tourism for the purpose of rural development. The focus of exploratory research is on gaining insights and familiarity for later investigation, or this is usually undertaken when problems are in a preliminary stage of investigation (Cuthill 2002). This study will provide future research implications for researchers and have managerial policy implications for development of rural tourism industry. Research design of the study includes phenomenology, case study, grounded theory, and content analysis. Phenomenology assists us to understand the meaning of people's lived experience. A phenomenological study explores what people experienced and focuses on their experience of phenomena (Groenewald 2004). A case study

S. Kumar, *Sustainable Rural Tourism in Himalayan Foothills*,
https://doi.org/10.1007/978-3-031-40098-8_3

Table 3.1 Research design of the study

Classification	Type of research design
Measurement of research	Applied
Category or purpose of research	Exploratory research
Research design	Phenomenology
	Case study
	Grounded theory
	Content analysis
Time dimension	Cross sectional studies (at point of time)
Sampling technique	Convenience sampling
	Judgment sampling
	Snowball sampling
	Quota sampling
Method of data collection	Observations
	Survey
	Secondary data
Statistical test used	Flexible strategies

Source: Author's elaboration

research design is one in which a particular object is studied as a subject matter and results obtained are generalized for the similar objects. The case method has its roots in sociology and has been found to be especially valuable in practice-oriented fields, such as management, public administration, psychology, history, education, and medicine (Mohajan 2018). Content analysis is a qualitative analytical technique used for analyzing secondary data such as textual materials, research papers, articles, newspaper reports, advertisements, and folklores (Cho and Lee 2014). The principal strength of grounded theory qualitative research is to discover, develop, or generate a theory from the observation and data collection from the research participants (Masoodi 2017), (Creswell 2013). The present study is cross-sectional in nature. A cross-sectional study has no time dimension and relies on existing differences rather than change following intervention, and groups are selected based on existing differences rather than random allocation (Hall 2008). The present study has been conducted once in a predetermined time frame. For the purpose of data collection, various nonprobability sampling techniques have been used. Nonprobability sampling techniques are those where the selection of participant is not on the basis of any mathematical calculation but on the attitude of the investigator. Convenience sampling is a technique where the selection of participants in a sample depends upon the occurrence of participants close to the observer (Setia 2016). Snowball sampling is used when it is difficult to find the reliable sources for data collection due to limited knowledge availability about the sampling population (Taherdoost 2016). Judgmental sampling is a sampling technique where the samples are selected on the basis of judgement of the investigator (Elfil and Negida 2017). Quota sampling is a technique for data collection where a fixed number of samples are obtained under various categories. These categories are divided on the basis of some characteristics. Samples with homogeneous characteristics are grouped together (Bhardwaj 2019). The data has been collected by conducting survey using two separate structured questionnaires. One questionnaire is used to collect data from the tourists, while the second questionnaire was used to collect data from other stakeholder that includes hoteliers, restaurant owners, photographers, travel agents, tourist guides, academicians, and policymakers. The data has also been collected from the published literature available on the theme of rural tourism. The study has sued flexible strategies for the purpose of data analysis. The strategies used are interpretive structure modeling (ISM), Matrice d'impacts croisés multiplication appliquée á un classement (MICMAC analysis), continuity-change matrix (C-C matrix), high-medium-low–vital-desirable-burden (HML-VDB) analysis, and destination discontinuity model.

3.2 Objectives of the Study

The following are the objectives of the study:

- To analyze the current status and patterns of rural tourism in India and particularly in Himachal Pradesh.
- To evaluate the consequences of rural tourism on socioeconomic structure of Himachal Pradesh.
- To identify the factors impacting a tourist decision to undertake rural tourism and develop hierarchical relationship among the factors.
- To formulate the framework for promoting financial inclusion and social inclusion with the help of sustainable rural tourism.
- To assess the role of technology in tourism industry in achieving the objective of social and financial inclusion.
- To formulate strategic crystal for development of sustainable rural tourism that will assist in transforming rural India.

3.3 Research Hypothesis

A scientific method to investigate a prediction or relationship hypothesis is referred to as research hypothesis. A research hypothesis helps to understand and construct tentative hypothesis on subject area of research study. Research hypothesis is a predictive assertion which relates independent variable to dependent variable. It should be clearly formulated, simple, and in declarative form. Moreover, it must be measurable and quantifiable so that statistical relationship can be ascertained.

According to Kerlinger (1986), "hypothesis is a conjectural statement of the relationship between two or more variables." However, Grinnell Jr. and Unrau (2010) stated that "a hypothesis is a written is written in such a way that it can be proven or disproven by valid and reliable data- it is in order to obtain these data that we perform our study."

For the study the following null hypothesis have been formulated:

- Ho_1: The level of awareness about rural tourism and its products is independent of tourist nationality.
- Ho_2: The level of awareness about rural tourism and its products is independent of tourist age.
- Ho_3: The level of awareness about rural tourism and its products is independent of type of stakeholders.
- Ho_4: Development of rural tourism is independent of use of technology by stakeholders.
- Ho_5: Development of rural tourism is independent of infrastructure development.
- Ho_6: Development of rural tourism is independent of growing environmental conscience among people.
- Ho_7: Development of rural tourism is independent of attitude of locals to adapt.
- Ho_8: Development of rural tourism is independent of travel motives of tourists.
- Ho_9: Development of rural is independent of marketing of destinations.
- Ho_{10}: Development of rural tourism is independent of support of local government and community.
- Ho_{11}: Development of rural tourism is independent of availability of funds with planning agencies.
- Ho_{12}: Development of rural tourism is independent of participation in private sector.
- Ho_{13}: Development of rural tourism is independent of destination characteristics.
- $Ho_{14:}$ Perception on rural tourism success is independent on the type of stakeholders.

3.4 Research Framework

A research framework is a blueprint that specifies the details of the procedures necessary for obtaining information for conducting the research projects. The objective of exploratory research design is to provide insights and understanding of situation, whereas conclusive research design is designed to assist the decision-maker in determining, evaluating, and selecting the best course of action to take in a given situation. The major objective of descriptive research design is to describe something—usually market characteristics or functions. It is also useful to obtain information regarding the current status of the phenomena and to describe "what exists" with respect to variables in a situation. Moreover, descriptive research design was chosen due to its suitability with objectives of this research study. The research design includes the following elements.

3.4.1 Study Area and Sampling Population

The present study uses case study model to develop a framework for sustainable rural tourism development. The research area selected for the study is Himachal

Pradesh, a state of India, located in the Himalayan Belt. It is surrounded by state of Jammu and Kashmir in north and north-west, Punjab to the south-west, Haryana and Uttar Pradesh to the south, and Uttarakhand to the southeast and is located at 31.1048° N, 77.1734° E. The state has 12 districts—Bilaspur, Mandi, Shimla, Kullu, Kangra, Sirmaur, Kinnaur, Lahaul and Spiti, Hamirpur, Una, Solan, and Chamba. Shimla is recognized as the capital of the state. The rationale behind selecting Himachal Pradesh for the study is its recent growth in tourism statistics and the threat of environment danger looming on it, especially since Himalayan Tsunami in Uttarakhand in year 2013. The study is conducted in all the 12 districts of Himachal Pradesh, and the sample size is based on the proportion of tourist visits in the state during 2018. In tune with the UN sustainable development goal no. 5, that is, mutual reinforcement of gender equality and sustainable development, the data has been collected from both male and female genders. The gender dimensions can be inextricably linked to tourism development across social, economic, and environmental domains. The incorporation of gender linkages to the rural tourism development sector shall embrace tangible ways forward and enhance toward accomplishment of the SDGs 2030 agenda. Literacy is another vital variable in the accomplishment of the country's travel industry. On the off chance that local people are not educated, then it will make a tremendous skill gap and could absolutely ruin the travel industry advancements. As per HPTDC report 2018, a total of 16,093,935 domestic tourists and 356,568 foreign tourists visited Himachal Pradesh in 2018–2019. As per HPTDC Report 2018, Himachal Pradesh has a total of 3349 guesthouses and hotels, 703 restaurants, and 1656 homestay units. It has a total of 2912 travel agencies, 899 photographers, 221 adventurers, and 1314 tourist guides.

3.4.2 Data Collection Tools

3.4.2.1 Data Sources

The techniques applied in the collection of data are determined by their effectiveness in addressing the aim and objectives of the study. For the purpose of the study, the data is collected from primary as well as secondary sources. For secondary data a document analysis has been conducted, on both printed and electronic materials relevant to the topic. The secondary sources are mainly used to study the current trends of tourism in the state. The policy analysis is conducted to identify the role and working of policymakers in respect of sustainable tourism in their area. The secondary data sources for the study includes:

- Central and state government reports and documents.
- Academic articles in periodicals.
- Research papers published.
- Reports of private agencies working in the related fields.
- Newspaper or website articles.

For the purpose of this study, a range of documents like yearly Tourism Statistics by Ministry of Tourism, Government of India, Tourism Statistics by Department of Tourism & Civil Aviation Himachal Pradesh, Tourism Policy 2018 by Himachal Pradesh Tourism Development Corporation, and socioeconomic reports of Himachal Pradesh were gathered and analyzed. Other records have been obtained from the government sites on the Internet and through multiple sources such as books, articles, conference papers, newspapers, magazines, maps, and various public records.

The primary data collection is conducted through fieldwork and will involve structured and semi-structured interviews, observation, and formal and informal discussions. For this a structural questionnaire will be prepared having a combination of open-ended and multiple-choice questions to be judged on a 7-point Likert scale. The following are the primary data sources considered for the study:

- Tourists (both males and females).
- Hoteliers and homestay owners.
- Restaurant operators.
- Travel agents in Himachal Pradesh.
- Government officials.
- Local population representatives.
- Researchers and academicians.
- Adventurers.
- Photographers.
- Tourist guides.

For the purpose of this study, structured and semi-structured interviews in a face-to-face situation and telephone interview and survey method is adopted to collect data. These interviews and surveys were used to collect information from central and local government officials, tourism operators, and representatives of local residents. These concentrated on tourism policies, the strategies, and plans for rural tourism development. The survey with the tourists focused on tourism experiences in Himachal Pradesh and their perception on rural tourism.

3.4.2.2 Questionnaire

Questionnaire acts as a means to collect information about people's knowledge, attitudes, behavior, and beliefs about a particular topic (Boynton and Greenhalgh 2004). Exploratory questions have been found to be successful in qualitative studies where the objective of collecting data is to gain insights on the topic (Syed Muhammad 2016). Therefore, several open-ended and objective questions were asked to the respondents to enquire about their perception on development of rural tourism industry. In addition, questions on general issues pertaining to tourism industry were also asked to stakeholders in tourism industry. For the present study, two separate questionnaires are prepared. The first is used to collect data from the sample drawn of the tourists, while the second is used to collect the data from the sample of other stakeholders. In the questionnaire of tourists, apart from collecting the demographic characteristics of the respondents, questions focused on their perception regarding the success or failure of rural tourism industry in Himachal Pradesh. Moreover, a set of questions were asked on their tourism behavior exhibited during their stay, the duration of stay, reasons for liking or disliking during stay, perception on impact of stay on environment, and issues faced by them during their stay. The questionnaire pertaining to other stakeholders focused is designed as per the various sub-categories of other stakeholders. Likewise, hoteliers and homestay owners are asked information about their industry, demands of tourist during their stay, occupancy rate of their hotels, use of technology by them during their stay, and their willingness to work in rural areas if the rural tourism industry is developed. Similarly, restaurant owners are asked about the behavior of tourist with respect to traditional food dishes. Travel agency operators are asked about their perception on willingness of tourists to visit rural areas in Himachal Pradesh if they are persuaded by offering economic incentives or through persuasive marketing schemes. Similar questions are asked from local cab drivers, photographers, and tourist guides on their customer base, impact of technology on their business, and their willingness to serve in rural areas of Himachal Pradesh if an opportunity arises. Lastly, academicians and local population representatives are asked about the contribution of the tourism industry in the economy of Himachal Pradesh, role of planning agencies in development of industry, strength and pit-falls for the industry, and their perception on pros and cons in spotlight of development of rural tourism industry in the state (Fig. 3.1).

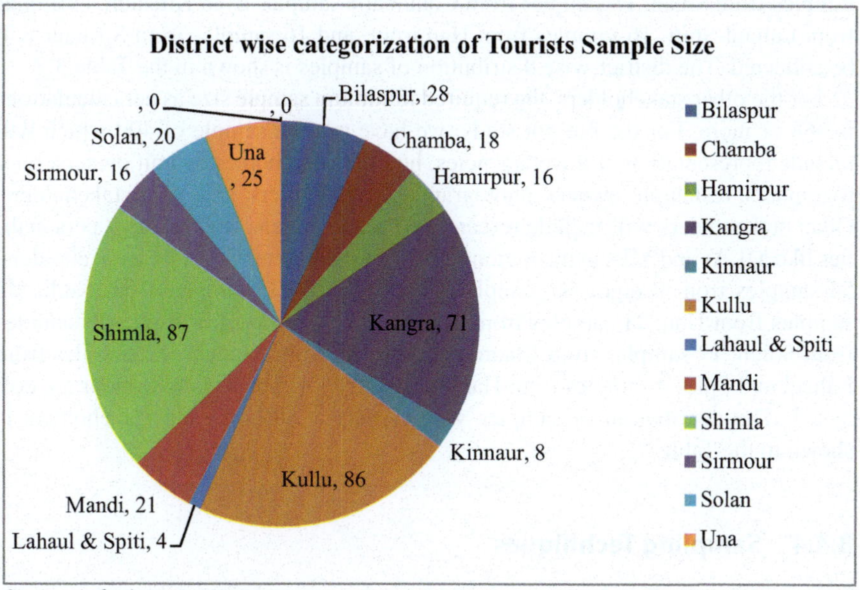

Source: Author's representation

Fig. 3.1 Distribution of tourist sample size. (Source: Author's representation)

Table 3.2 Tourist sample size (on the basis proportion of tourists who visited H.P. in 2018)

Sr. No.	District	Domestic tourist	Foreign tourist	District total
1.	Bilaspur	28	0	28
2.	Chamba	18	0	18
3.	Hamirpur	16	0	16
4.	Kangra	42	29	71
5.	Kinnaur	7	1	8
6.	Kullu	58	28	86
7.	Lahaul & Spiti	1	3	4
8.	Mandi	19	2	21
9.	Shimla	52	35	87
10.	Sirmaur	15	1	16
11.	Solan	19	1	20
12.	Una	25	0	25
Total		**300**	**100**	**400**

Source: Author's compilation

3.4.3 Sampling Size

The sample size for tourists calculated for 95% confidence level with 5% margin of error is 385 or more. So, for the study a sample of 400 tourists is taken with 300 domestic tourists and 100 foreign tourists. On the basis of proportion of visits in the year 2018, 71 samples from Kangra, 87 samples from Shimla, 86 samples from Kullu, 25 samples from Una, 21 samples from Mandi, 28 samples from Bilaspur, 20 samples from Solan, 18 samples from Chamba, 8 samples from Kinnaur, 4 samples from Lahaul-Spiti, 16 samples from Hamirpur, and 16 samples from Sirmaur will be collected. The district-wise distribution of samples is shown in the Table 3.2.

For the other stakeholders, the required minimum sample size as per calculations is 368 or more. For the present study, we have taken a sample of 400 which will include representation of travel agencies, hoteliers and homestay unit owners, photographers, restaurant owners, photographers, adventurers, and other stakeholders. Other stakeholders will include researchers, academicians, and eminent personalities like MLAs and MPs in the region. On the basis of number of other stakeholders, 53 samples from Kangra, 97 samples from Shimla, 137 samples from Kullu, 05 samples from Una, 21 samples from Mandi, 7 samples from Bilaspur, 29 samples from Solan, 19 samples from Chamba, 6 samples from Kinnaur, 12 samples from Lahaul and Spiti, 4 samples from Hamirpur, and 11 samples from Sirmaur are collected. The distribution of samples with respect to districts and stakeholders is shown in the Table 3.3.

3.4.4 Sampling Techniques

For selection of sample of tourists, convenience and judgmental sampling techniques are used. The purpose of using convenience sampling is based on the reason

Table 3.3 Other stakeholders' sample size (on the basis of registered stakeholders as per HPTDC, Reports 2018)

Sr. no.	District	Hotels and guesthouse	Homestay units	Tourist guides	Photographers + adventurers	Travel agencies	Cab/taxi operators	Restaurant	Other stakeholders	District total
1.	Bilaspur	3	1	1	0	0	0	1	1	7
2.	Chamba	7	3	4	1	1	1	1	1	19
3.	Hamirpur	2	1	0	0	0	0	1	0	4
4.	Kangra	21	6	5	0	7	6	4	4	53
5.	Kinnaur	2	2	0	0	0	0	0	1	6
6.	Kullu	29	11	12	28	23	21	4	9	137
7.	Lahaul and Spiti	3	5	1	0	1	1	1	1	12
8.	Mandi	7	4	3	0	2	2	2	2	21
9.	Shimla	19	10	7	14	19	17	5	6	97
10.	Sirmaur	4	1	2	0	0	0	3	2	11
11.	Solan	11	3	3	0	2	2	6	3	29
12.	Una	4	0	0	0	0	0	1	0	5
Total	111	46	38	43	56	50	27	30	400	

Source: Author's compilation

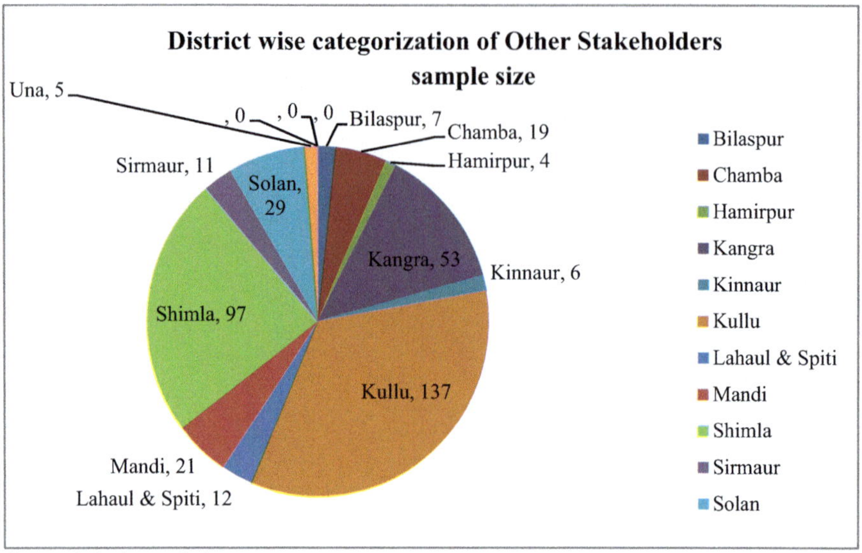

Source: Author's compilation

Fig. 3.2 Distribution of other stakeholders' sample size. (Source: Author's compilation)

that the participants were selected exclusively on the logic of their presence in proximity to the investigator. The tourists were selected as per the convenience of the investigator because there is no risk of any bias in the sample selection. For selection of sample of other stakeholders, first, judgmental sampling technique is employed. Stakeholders were recognized through professional contacts in government departments, state tourism development corporation, relevant authorities, as well as the network of individuals and organizations that were involved in tourism on each of the islands. Further, stakeholders were selected using snowball sampling technique. Snowball sampling is helpful as the experts and other scholars researching in the area could guide more about who would better throw light on the topic. The use of quota sampling has been done in the study as there is limit on the number of different other stakeholders that were targeted for the data collection (Fig. 3.2).

3.5 Data Analysis Tools and Techniques

Data analysis means to analyze the data to achieve significant conclusions. Research cannot produce results without effective data analysis. Data analysis can be quantitative as well as qualitative. Wherein quantitative analysis deals majorly with numbers, qualitative analysis is concerned with answering "What," "How," and "Why" aspects of the statement of problem. This is a qualitative study and makes use of the following techniques for the purpose of data analysis.

3.5.1 Interpretive Structure Modeling (ISM)

Interpretive structural modeling (ISM) developed from the works of John Warfield in the late 1960s. This method is based on the judgment of a group of people on how the things are related and based on it provides a structural relationship. This hierarchical structure is important because direct and indirect relationships developed through it provide a holistic view of the working of the system. In order to develop a strategy, one must know about key areas that generate concerns and areas that could prove to be significantly beneficial. ISM helps in the identification of such areas. The researchers across all the fields are using ISM to chalk out the flexible strategies since long. Mandal and Deshmukh (1994) used ISM to determine the basis for selection of vendor by the purchasing department. Raj et al. (2008) used ISM to identify and rank the enablers of a flexible manufacturing system in India. Rizvi et al. (2019) identified the determinants of nonperforming assets in the banking industry and studied the interrelationship among them. ISM is considered as interpretive as the analysis is based on the interpretation of a group of people. Similarly, it is a modeling technique as the result is represented in a form of digraph model (Raj et al. 2008).

Step 1: The first step in the process is the identification of factors influencing the system. The factors could be identified from the literature review or by use of the Delphi technique. The factors and the interrelationship could then be verified using expert opinion.

Step 2: The next step is the formulation of a self-structuring interpretive matrix. This matrix represents the interrelationship between the factors in the form of V-A-X-O coding.

Step 3: The SSIM matrix is then converted in the binary form of zeroes and ones. The resultant matrix is known as initial reachability matrix.

Step 4: The initial reachability matrix is then transformed into the final reachability matrix by incorporating the rule of transitivity in the interrelationship between the factors. This step also includes determining the driving power and dependence level of each of the factor involved. The calculation then helps in the classification of factors into categories of a driving factor, dependent factor, autonomous factor, and linkage factor.

Step 5: The step that follows is calculating the partition level of each of the factor in the final ISM digraph. The level is calculated using the reachability set, antecedent set, and intersection set and a series of iterations.

Step 6: The final step is the formulation of ISM digraph indicating the links between all the links. The transitive links are then removed to give the relationship between the factors a hierarchical structure. The following figure presents the generalized framework for interpretive structure modeling to be used in present study (Fig. 3.3).

As identified from the literature, the policies developed for the development of rural tourism have been rigid in nature (Kumar et al. 2019). They are not able to

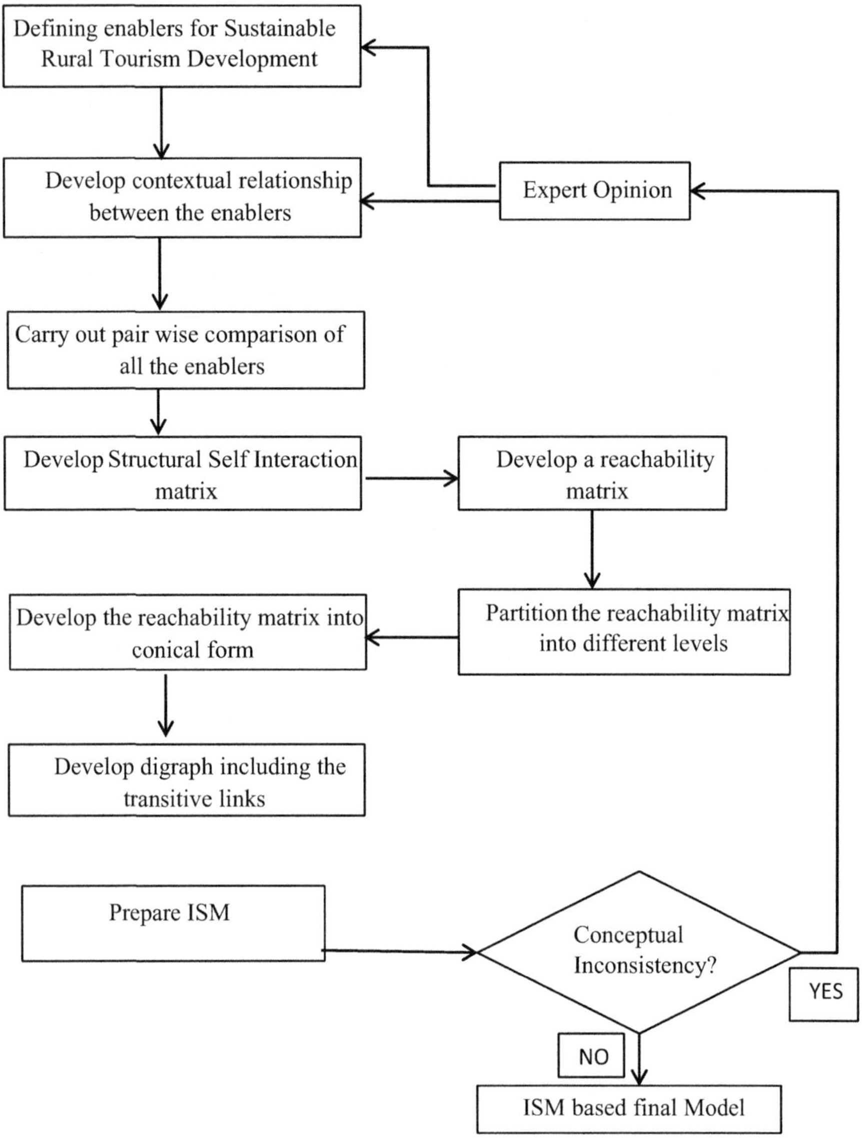

Source: Author's compilation

Fig. 3.3 Research methodology for ISM. (Source: Author's compilation)

adapt themselves to the dynamic environment without any outside stimulus. This results in frequent overhauling of strategies developed. As this is a time-consuming process, policymakers are reluctant to chalk out strategies from the scratch. This has hampered the development of rural tourism industry. On the other hand, strategies formulated using flexible strategic tool such as interpretive structure modeling results into strategies that are flexible in nature and eliminate the need to overhaul them due to change in the environment (Kulkarni et al. 2018). For preparation of flexible strategies, interpretive structure modeling (ISM) will be used. With the help of ISM, factors affecting sustainable tourism development will be identified and ranked. The drivers and dependent factors will be determined, and strategies will be formulated so as to reduce the risk. New products can be established by identifying the hierarchy in which factors affect new product development.

3.5.2 MICMAC Analysis

MICMAC is a cross-impact matrix multiplication applied to classification as a structural prospective analysis used to study indirect relationships. MICMAC analysis involves the development of a graph that classifies factors based on driving power and dependence power. MICMAC analysis is used to classify the factors and validate the interpretive structural model factors in the study to reach desired results and conclusions. The graph for MICMAC analysis looks like Fig. 3.4, wherein the dependence level of a variable is placed on X axis and driving power of the variable is placed on Y axis.

This analysis helps in classification of variables into four categories:

- Driving factors: These are the factors that have high degree of driving power and a low degree of dependence level.
- Dependence factors: These factors are characterized by high degree of dependence level and a low degree of driving power.
- Autonomous factors: These factors have low degree of driving power as well as dependence level.

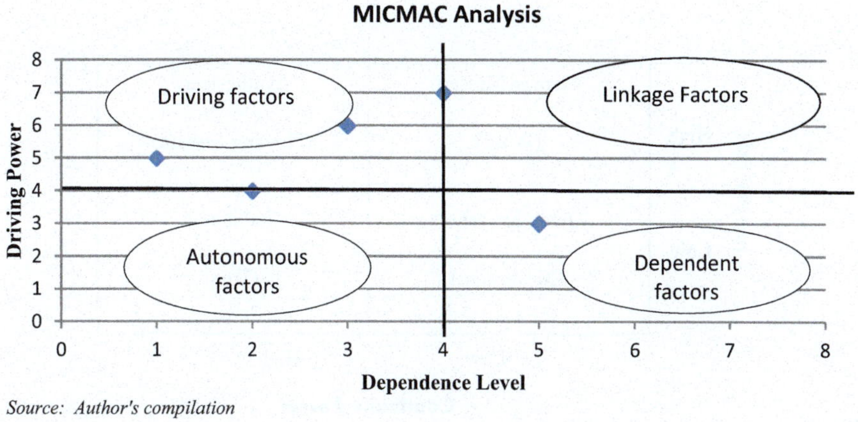

Fig. 3.4 Generalized model for MICMAC analysis. (Source: Author's compilation)

- Linkage factors: These are the factors with high degree of dependence level as well as driving power.

While formulating flexible strategies, it is very important to know the classification of variable. Since an autonomous factor is beyond the control of the policymakers, nothing much can be done to control it. As driving factors influence many factors in the system, the need to focus on such factors while framing policies and strategies is significant. Linkage factors transfer the impact of driving factors onto the dependent factors. And the steps taken for the driving factors should be taken after considering their impact on the dependent factors.

3.5.3 Continuity-Change Matrix

To better understand the tourism industry, a continuity-change analysis will be performed. A continuity-change matrix would be used to develop the future course of action for the industry. The process of C-C matrix includes:

Step 1: Identify the different continuity and change factors for the tourism industry in Himachal Pradesh. *Continuity forces* are the ones which persuade the entity to keep moving in the direction of its strategies. *Change forces* on the other hand are factors motivating to shift the focus toward new opportunities. A continuity-change matrix is derived from the interaction of these two factors (Fig. 3.5).

On conducting the analysis, the appropriate quadrant for the tourism industry will be determined, and strategies would be formulated. The meaning of each of the quadrant is as follows:

- High continuity force and low change force: These industries/organizations in this category are called as "stabilizers" and typified as "tree," as these are usually more stable like a tree and slowly evolve overtime maintaining continuity in their strategy.

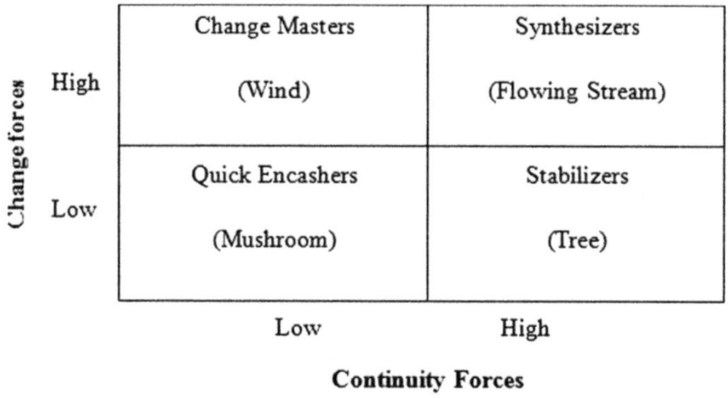

Fig. 3.5 A continuity-change matrix model. (Source: Author's compilation)

- Low continuity force and low change force: The industries/organizations in this category have low bindings on either side and can be treated as "quick encashers" typified as "mushroom." These organizations can quickly encash the opportunities and can shift to new ones due to low continuity forces.
- Low continuity force and high change force: These are "change masters" typified as "wind," as these industries/organizations have low pull due to existing framework and need to be changed radically due to new opportunities or innovative technological change.
- High continuity force and high change force: These industries/organizations need to deal with a confluence of continuity and change and are the "synthesizers" metaphorically called as "flowing stream." The concern of simultaneous continuity and change is maximum under this category. Flowing stream strategy analyzes all the forces and propose different changes, shifts, and partitions depending upon the intensity of factors.

The purpose of this analysis is to identify different pathways for rural tourism industry to follow. Since there is no one definite route that could ensure the survival of the rural tourism industry in long run, this analysis will help us to determine the different pathways that the industry could move to depending upon its own strengths, weaknesses, opportunities, and threats in the environment.

3.5.4 HML-VDB Analysis

The continuity factors are classified into vital (factors that are one's key strengths and are absolutely necessary for survival), desirable (factors that are important for generating revenue, but are not necessity), and burden (factors that are existing due to one's own limitations) factors. The change factors are classified into high (opportunities and threats that can have severe impact on survival), moderate (environmental factors that have moderate impact on survival of industry), and low (factors with relatively negligible impact on survival of industry) impact factor (Fig. 3.6).

The strategies are prepared ignoring the factors which fall under burden and have moderate or low impact on the industry. This analysis will be useful in determining

	High Impact	Moderate Impact	Low Impact
Vital			
Desirable			
Burden			

Source: Author's compilation

Fig. 3.6 A generalized framework for HML-VDB analysis. (Source: Author's compilation)

the pathways using C-C matrix by eliminating the probable pathways that are either not feasible or will not generate significant future gains to the industry.

3.5.5 Destination Discontinuity Model

Destination discontinuity model is utilized to shift the burden of tourists from exhausted places to such underexplored places. The model will also be used to determine the economic impact of the move by analyzing the performance of new destination in comparison to the previous destination.

On X axis there is a time frame which shows the time period of growth for a destination. On Y axis it is performance of the destination. The destination over a period of time performs like a product behaves during product life cycle. When destination reaches to its saturation level, the destination starts losing more growth options, and tourist inflow cannot be increased further. This is the time when a destination should be discontinued, and efforts must be made for sustainable development of the destination while simultaneously searching for new tourism destination.

3.5.6 Tourism Typology Life Cycle

Every tourist destination passes through different stages in its development (Weiermair et al. 2015). Similarly, each tourism typology also passes through various stages during its life time. There are occasions where new tourism typologies are developed (Nordin 2005) and demand for certain tourism typology stagnates or completely declines. Different stakeholders in tourism industry therefore play a vital role in each of the stage of the tourism typology. The various stages through which a tourism typology passes through are:

- Exploration: Under this stage, a new tourism typology is identified and starts to develop. Strategies are framed to make tourists aware about the typology and invite them. Fundamental framework is provided at the destinations offering the tourism typology.
- Involvement: As the awareness about the typology increases, the number of tourist consuming the product also expands. Local people also start consuming the tourism product.
- Development: In this stage, the popularity of the tourism typology has soared due to the marketing efforts by the planning authorities and stakeholders. Infrastructure and relevant market have been developed for the particular typology.
- Consolidation: In this stage, the share of the tourism typology increases, and it becomes a prominent tourism product offered. Marketing and advertising efforts are further wide-reaching to extend the tourist season and market area.

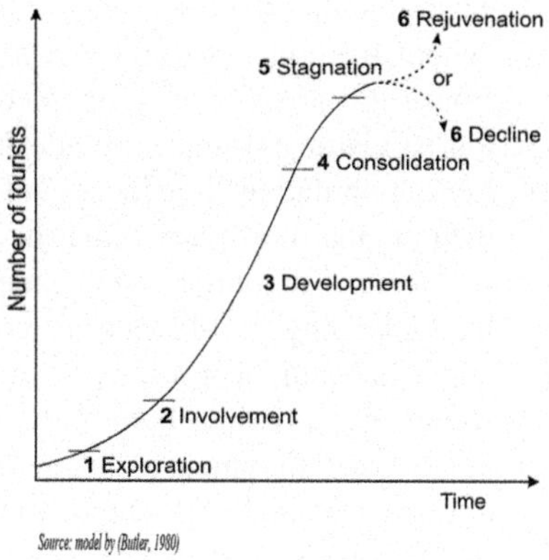

Fig. 3.7 Tourism typology life cycle. (Source: Model by (Butler 1980))

- Stagnation: In this stage, the typology attracts tourists at a normal rate but has lost its charm due to prolonged over tourism. The environmental, social, and economic damages due to the tourism typology start to become visible.
- Rejuvenation or decline: In this stage, either the planning agencies completely abort the tourism typology due to its negative consequences or try to reimplement the tourism typology with substantial change in the product offering (Fig. 3.7).

This analysis is significant as under each of the stage different marketing strategies is adopted to make the typology still relevant for reaping economic gains. The stage for rural tourism industry in Himachal Pradesh is assessed, and relevant marketing mix strategies are recommended for the development of industry.

3.6 Research Implications

The study will have implications for every stakeholder involved in tourism industry. The following section highlights the research implications from the perspective of different stakeholders:

- Government and planning agencies: The government could accomplish a few of its goals through this study. First, this is in accordance with objective of sustainability in which the government is practicing in recent times. Second, financial inclusion and social inclusion can be achieved more effectively in rural areas through tourism. Third, generating skill-based employment opportunities has been a challenging role for governments in recent times (Saini 2015). Thus, the study would open doors for missions such as Skill India to go and target rural population and enable them to serve as stakeholders in tourism industry. Fourth, assuming by one way or another private sector is attracted to work for rural tourism development, it would save huge amount of government money which could be utilized somewhere else (Kaurav et al. 2013). Fifth, government could make economically weaker section of the society self-reliant. Sixth, government could benefit from increase in tax revenue as vast majority of the agriculture income from rural areas is tax exempted. Seventh, a dataset can be prepared for rural tourist visiting Himachal Pradesh.

- Local population: For locals, infrastructure developed at destination would lift their standard of living (Atsushi 2011). They would find it much easier to communicate with tourist and outside world and would be able to better know about the happenings worldwide. It would be a platform for them to show their local talent and attract millions to their social values customs and traditions. This would also provide them with an alternate source of earning. High-scale disguised unemployment problem in rural area could be tackled by developing rural tourism sustainably (Shekhar and Attri 2017). By creating employment opportunities in their own backyard, migration to urban areas for search of livelihood could be restricted.
- Tourists: For tourist, a new tourism product would be developed which would be new and unexplored. This creates an excitement and an altogether new reason to visit a destination (Praveen kumar 2015).

3.7 Methodological Limitations

In spite of the inherent merits attributed to the utilization of questionnaires for the purpose of data generation, empirical observations underscore the necessity for adopting blended methodologies that integrate both qualitative and quantitative techniques. The quantitative and subjective approaches create various outcomes and as such are complementary. The study has few constraints that exist in the context of qualitative study. Huberman and Miles (1994) acknowledged that in qualitative research in the social sciences, there may exist a number of issues such as labor-intensiveness of (qualitative) data collection, the possibility of existence of researcher bias, issues in handling the data, the time constraint in processing and coding data, the issue in generalizability of findings, the statistical evidence of conclusions, and the managerial and policy implications arising from the study. This present study acknowledges the following limitations in its adopted methodology:

- The documents analyzed for the study were searched over the Internet and other published sources. There is a chance that there was only a limited or partial set of documents that were analyzed. Besides, the analyzed documents may have been written explicitly for another tourism typology, and it is very difficult to evaluate causal relationship simply through furthering how and why questions beyond the scope of the document's intent.
- The sample size for the field work is limited due to the financial and time constraint for the study. In addition, the lockdown imposed during the study period due to the COVID-19 pandemic likewise made it difficult to cover a huge sample of data.

- This qualitative study is based on the perception of tourists and a variety of other stakeholders. Albeit essential precautionary measures have been taken to eliminate the bias in the study, quite possibly the insight may vary if a different and particular arrangement of interviewees was chosen.

References

Atsushi, I. 2011. Effects of improving infrastructure quality on business costs: Evidence from firm-level data in eastern europe and central asia. *The Developing Economies* 49 (02): 121–147.

Bhardwaj, P. 2019. Types of sampling in research. *Journal of the Practice of Cardiovascular Science* 5: 157–163.

Boynton, P.M., and T. Greenhalgh. 2004. Selecting, designing, and developing your questionnaire. *BMJ* 328 (7451): 1312–1315. https://doi.org/10.1136/bmj.328.7451.1312.

Butler, R.W. 1980. The concept of a tourist area cycle of evolution: Implications for management of resources. *Canadian Geographer/ Le Géographe Canadien* 24 (01): 5–12.

Cho, J.Y., and E.H. Lee. 2014. Reducing confusion about grounded theory and qualitative content analysis: Similarities and differences. *The Qualitative Report* 19 (32): 1–20.

Creswell, J.W. 2013. *Qualitative inquiry research design, choosing among five approaches*. Los Angeles: Sage.

Cuthill, M. 2002. Exploratory research: Citizen participation, local government, and sustainable development in Australia. *Sustainable Development* 10: 79–89.

Elfil, M., and A. Negida. 2017. Sampling methods in clinical research; an educational review. *Emergency(Tehran)* 5 (01): 52.

Groenewald, T. 2004. A phenomenological research design illustrated. *International Journal of Qualitative Methods* 3 (01): 42–51.

Grinnell Jr, R. M., & Unrau, Y. A. 2010. *Social work research and evaluation: Foundations of evidence-based practice*. Oxford University Press.

Hall, J. 2008. Cross-sectional survey design. In *Encyclopedia of survey research methods*, ed. P.J. Lavrakas, 173–174. Thousand Oaks: Sage.

Huberman, A.M., and M.B. Miles. 1994. *Qualitative data analysis: An expanded*. 2nd ed. Thousand Oaks: Sage.

Kerlinger, F. N. 1986. *Foundations of Behavioural Research* (3rd edn). New York: CBS College Publishing.

Kaurav, R.P.S., J. Kaur, and K. Singh. 2013. Rural tourism: Impact study—an integrated way of development of tourism for India. In *Changing paradigms of rural management*, ed. R.K. Miryala, 313–320. Hyderabad: Zenon Academic Publishing.

Kulkarni, Prashant B., K. Ravi, and S.B. Patil. 2018. Interpretive structural modeling (ISM) for implementation of green supply chain management in construction sector within Maharashtra. *International Research Journal of Engineering and Technology (IRJET)*: 2460–2472.

Kumar, Suneel, Navneet Guleria Shekhar, and N. Guleria. 2019. Understanding dynamics of niche tourism consumption through interpretive structure modeling. *Saaransh RKG Journal of Management* 11 (01): 40–48.

Mandal, A., and S.G. Deshmukh. 1994. Vendor selection using Interpretive Structural Modelling (ISM). *International Journal of Operations & Production Management* 14 (06): 52–59.

Masoodi, M. 2017. A comparative analysis of two qualitative methods: deciding between grounded theory and phenomenology for your research. *Vocational Training: Research and Realities* 28 (01): 23–40.

de Mello, A.M., and M. Pedroso. 2018. Applied research articles: Narrowing the gap between research and organizations. *Revista de Gestão* 25 (04): 338–339.

Mohajan, H.K. 2018. Qualitative research methodology in social sciences and related subjects. *Journal of Economic Development, Environment and People* 7 (01): 23–48.
Nordin, S. 2005. *Tourism of tomorrow: Travel trends and forces of change*. European Tourism Research Institute.
Praveenkumar, S. 2015. Tourism marketing and consumer behaviour. *Research Journal of Social Science and Management* 4 (12): 73–81.
Raj, T., Shankar, R., & Suhaib, M. 2008. An ISM approach for modelling the enablers of flexible manufacturing system: The case for India. *International Journal of Production Research* 46 (24): 6883–6912.
Rizvi, N.U., S. Kashiramka, S. Singh, and Sushil. 2019. A hierarchical model of the determinants of non-performing assets in banks: An ISM and MICMAC approach. *Applied Economics*: 1–21.
Saini, V. 2015. Skill development in India: Need, challenges and ways forward. *Abhinav National Monthly Refereed Journal of Research in Arts & Education* 4 (04): 1–9.
Setia, M.S. 2016. Methodology series module 5: sampling strategies. *Indian Journal of Dermatology* 61 (05): 505–509.
Shekhar, Suneel, and K. Attri. 2017. Incredible India: SWOT analysis of tourism sector. In *Development aspects in tourism and hospitality sector*, 175–189. New Delhi: Bharti Publications.
Syed Muhammad, S. K. 2016. *Basic guidelines for research: An introductory approach or all disciplines*, 1st ed. Bangladesh: Book Zone Publication.
Taherdoost, H. 2016. Sampling methods in research methodology; how to choose a sampling technique for research. *International Journal of Academic Research in Management* 5 (02): 18–27.
Weiermair, K., M. Peters, and M. Schuckert. 2015. Destination development and the tourist life-cycle: Implications for entrepreneurship in alpine tourism. *Tourism Recreation Research* 32: 83–93.

Link to SDGs

THE 17 GOALS | Sustainable Development (un.org)

Rural Himachal Pradesh: A Strategic Approach for Solving Socioeconomic Challenges

4

4.1 Demographic Characteristics of Himachal Pradesh

Demographics can affect the behavior of the individuals; hence, it is essential to understand the demographic structure of the rural population in HP. The analysis will give insight into the concerns for the state and how the proposed study will be influenced by the demographics. For the study purpose, a close observation on the population, gender-wise composition, literacy rate, and sex ratio of the state shall be made. This analysis will also be conducted at the district and at the rural levels for more accurate analysis information.

4.1.1 Population

As per the census of India (2011) report, it was estimated that about 89.9% of the population still resides in rural areas in Himachal Pradesh. The total rural population for the state is 6,176,050 that includes 3,110,345 males and 3,065,705 females. Talking about the districts, then, Kangra (23.1%) has the highest proportion of rural population, followed by Mandi (15.2%), Shimla (9.9%), Chamba (7.8%), Shimla, Sirmaur, and Una (7.7% each), Hamirpur (6.9%), Kullu (6.4%), Bilaspur (5.8%), Kinnaur (1.4%), and Lahaul and Spiti (0.5%). The table below shows the district-wise distribution of the rural population in Himachal Pradesh (Table 4.1).

Talking about the proportion of males and females in the rural areas, the proportion of rural males to females is almost equal. The population of males and females is distributed as per the total population. Kangra district (22.7%) accounts for most males in rural areas. It is followed by Mandi (15.0%), Shimla (10.1%), Solan (8.0%), Sirmaur (7.9%), Chamba (7.8%), Una (7.7%), Hamirpur and Kullu (6.5%), Bilaspur (5.8%), and Lahaul and Spiti (0.5%). A similar spread is seen in the female population, with Kangra accounting for 23.4% of the rural female population. It is followed by Mandi (15.4%), Shimla (9.7%), Chamba (7.9%), Una (7.7%), Solan

S. Kumar, *Sustainable Rural Tourism in Himalayan Foothills*,
https://doi.org/10.1007/978-3-031-40098-8_4

Table 4.1 District-wise population in the rural areas of Himachal Pradesh (as per census 2011)

S. No	District	Total rural population					
		Male	% male pop.	Female	% female pop	Total	% of population
1.	Bilaspur	179,653	5.8%	177,174	5.8%	356,827	5.8%
2.	Chamba	241,963	7.8%	241,009	7.9%	482,972	7.8%
3.	Hamirpur	200,748	6.5%	222,590	7.3%	423,338	6.9%
4.	Kangra	705,365	22.7%	718,429	23.4%	1,423,794	23.1%
5.	Kinnaur	46,249	1.5%	37,872	1.2%	84,121	1.4%
6.	Kullu	203,269	6.5%	193,243	6.3%	396,512	6.4%
7.	Lahaul & Spiti	16,588	0.5%	14,976	0.5%	31,564	0.5%
8.	Mandi	466,050	15.0%	471,090	15.4%	937,140	15.2%
9.	Shimla	314,295	10.1%	298,364	9.7%	612,659	9.9%
10.	Sirmaur	246,175	7.9%	226,515	7.4%	472,690	7.7%
11.	Solan	249,736	8.0%	228,437	7.5%	478,173	7.7%
12.	Una	240,254	7.7%	236,006	7.7%	476,260	7.7%
	Total	**3,110,345**		**3,065,705**		**6,176,050**	

Source: Author's compilation

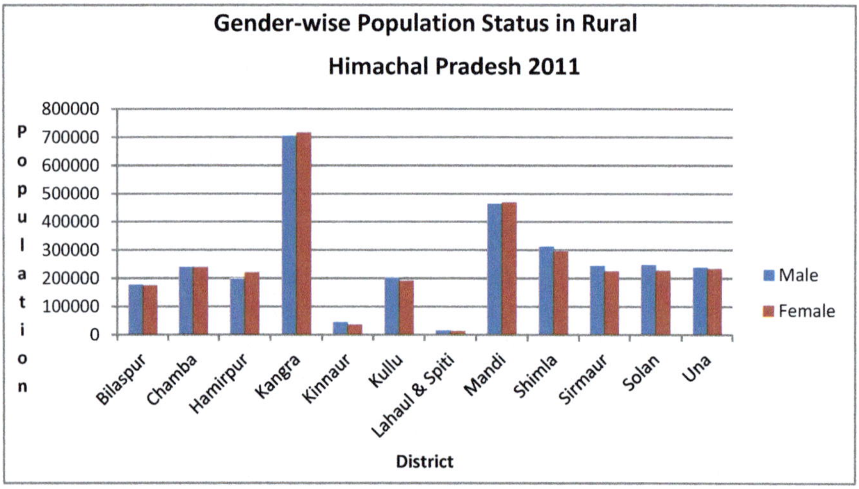

Source: Author's representation

Fig. 4.1 Gender-wise rural population of Himachal Pradesh. (Source: Author's representation)

(7.5%), Sirmaur (7.4%), Hamirpur (7.3%), Kullu (6.3%), Bilaspur (5.8%), Kinnaur (1.25), and Lahaul and Spiti (1.5%).

The results also indicate that the population is spread throughout Himachal Pradesh, and there will be no shortage of supply for the tourism industry. There is an available workforce for both the genders and across age groups as well (Fig. 4.1).

Barring two districts, i.e., Kinnaur and Lahaul & Spiti, the figure shows that human resource is abundant in rural areas. If we compare the statistics with the tourist inflow, in that case, the number of tourists visiting these two districts is not up to mark. This might be due to the inhabitable climate or lack of infrastructure to support lifestyle. Overall, rural Himachal Pradesh has enough human resources to provide to the rural tourism industry.

4.1.2 Literacy Rate and Sex Ratio

According to the definition by UNESCO, literacy rate is "the percentage of the population of a given age group that can read and write." A high literacy rate suggests an adequate primary education system and literacy programs that have enabled a large number of population to acquire the ability to use the written word (and making simple arithmetic calculations) in daily life to continue learning.

Sex ratio is defined as the number of females per thousand males at a point in time. This ratio indicates the society's attitude toward females and highlights social evils such as female infanticide (Table 4.2).

The literacy rate in the rural parts of the state is 81.9%. The literacy rate for males is 89.1% and for females is 74.6%. The results are encouraging but can be further improved. The literacy rate will have a social impact and can influence the development of the tourism industry. Education is often linked to the standards of living and human development. An educated person can be easily trained, can identify opportunities, and derive economic benefits. In addition, an educated person can be taught about behavior that they should exhibit with the tourists. An industry would require smart and educated people to work. Therefore, increasing literacy in the rural population will benefit the tourism industry in the long run (Fig. 4.2).

Table 4.2 Population demographics of rural Himachal Pradesh

S. No	District	Literacy rate			Sex ratio
		Male	Female	Total	Rural HP
1.	Bilaspur	90.9	77.2	84.1	986
2.	Chamba	81.5	59.7	70.6	996
3.	Hamirpur	94.3	82.1	87.8	1109
4.	Kangra	91.4	79.6	85.4	1019
5.	Kinnaur	87.3	71	80	819
6.	Kullu	87	69.5	78.5	951
7.	Lahaul and Spiti	85.7	66.8	76.8	903
8.	Mandi	89.2	72.6	80.8	1011
9.	Shimla	87.8	72.7	80.5	949
10.	Sirmaur	84.6	69.3	77.3	920
11.	Solan	88.6	75	82.2	915
12.	Una	92.1	80.9	86.5	982
	Total	**89.1**	**74.6**	**81.9**	**986**

Source: Author's compilation

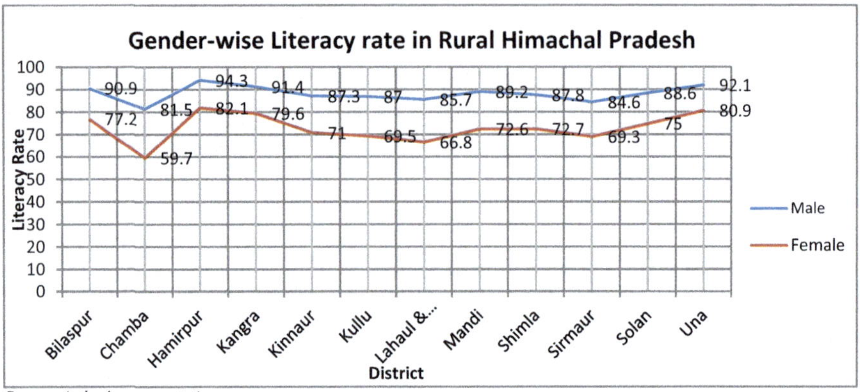

Source: Author's representation

Fig. 4.2 Gender-wise literacy rate in rural Himachal Pradesh. (Source: Author's representation)

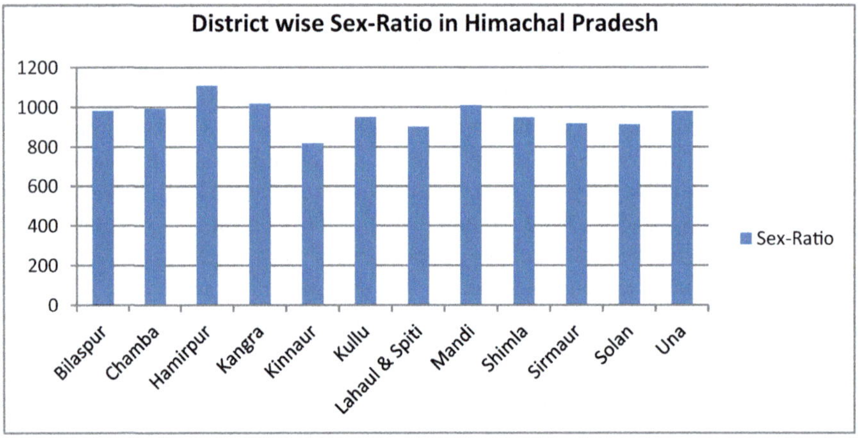

Source: Author's compilation

Fig. 4.3 District-wise sex ratio in Himachal Pradesh. (Source: Author's compilation)

The sex ratio in Himachal Pradesh is 972, i.e., it indicates 972 females per thousand males. In urban areas, this ratio dips to 853, while in the rural areas, the ratio rises to 986. The ratio is above the national ratio, which is 933. Interestingly, there are more females than per thousand males in three districts, i.e., Kangra, Hamirpur, and Mandi. This ratio is essential to understand the social issues prevailing in the society. However, it is necessary for developing the tourism industry. The presence of females in the rural society indicates the welcoming attitude for the females by the people. This means that there will be safety for female tourists in the rural regions. Similarly, the females can effectively address the needs and wants of female tourists (Fig. 4.3).

The conclusion from the above two graphs is that in the districts Kinnaur, Sirmaur, Solan, and Lahaul and Spiti, where the literacy level among females is low, the sex ratio is also low. This indicates that literacy in females is essential to uproot the social evils from the society.

4.2 Economic Challenges for Himachal Pradesh

HP is one of the most prosperous states in India. The unemployment rate in HP is around 8.06%. The people's per capita income is Rs.195,255 in 2019–2020, registering a growth rate of 6.6% over the previous fiscal year. However, there exist several challenges in the rural areas of HP that hinder its economic development. These challenges need to be addressed in the present times to make the state more economically developed. Some of these challenges are as follows:

- Dependence on agriculture for employment: The rural HP is dependent on agriculture for its employment and income generation. The primary activity for rural people is agriculture, horticulture, and livestock rearing. This dependence on agriculture affects their income because of the industry's shortcomings. Thus, generating more employment opportunities and a regular income source is an uphill task for rural Himachal Pradesh.
- Less income-earning prospects in the state: The absence of manufacturing industries and a flourishing service sector causes low income-earning opportunities in the rural areas. The lack of infrastructure also prohibits activities that can act as an alternate income source for the people.
- The tourism industry's weak performance: The tourism industry is far more developed in urban regions than in rural areas. Due to lack of infrastructure such as communication and accessibility, poor hospitality conditions, and low demand, the tourism industry in rural areas has shown lackluster performance. The tourism products offered in rural areas are not fully developed and are not marketed effectively.
- Lack of infrastructure in rural areas: Poor living standards in rural areas make it noneconomical for the government to develop rural areas' infrastructure. In the absence of infrastructure, people are unwilling to demand products and services necessary for their upliftment. Due to the lack of essential resources, rural people do not use financial services.
- Low financial inclusion: The level of financial inclusion in the rural HP is also not encouraging. There are bank accounts of people, but the numbers of operative bank accounts are less. Other financial services such as digital payment, investing in bank deposits, and availing other credit schemes are also less in the state's rural part.

4.3 Social Challenges for Himachal Pradesh

A society's future is marred with the presence of social evils and challenges. Society generates development when it relinquishes itself from its past chains and move toward its upliftment. However, there are challenges in its way. These challenges may be inherent customs and social values, or they might be due to their low education level. Nevertheless, these challenges limit the development of humans and, therefore, must be addressed carefully.

- Low skill and technical development in youth: The lack of job requiring skill development and nonavailability of skill enhancement measures result in the youth less skill-oriented in the rural HP. The majority of jobs in rural areas are in agriculture, an industry where the quantity of labor is preferred over the skills. Therefore, the skill development in the rural areas is deficient in youth. There is a lack of innovation and creativity in people. This might hinder the mental growth and development.
- Low education among females: The literacy rate among females in the rural areas is 74.6. This percentage is relatively low as the number of females in the society is almost equal to the male population. An educated female not only educates her family but also takes proper care of hygiene and health. Female education is an essential aspect of the social development of the society.
- The low standard of living in rural areas: Since the income and education level in the rural areas is low, the rural people's standard of living is not up to the mark. The absence of knowledge about modern technology causes a lack of demand for products and services. Thus, the standard of living in the rural HP is far below the urban standards.
- Low confidence and motivation level in the youth: In the absence of opportunities for work and rewards for innovation and creativity, the youth finds almost no reason to work more efficiently and effectively. Thus, the motivation level among youth is relatively low in the state. Further, because of low infrastructure development and low living standards, the youth lack confidence in dealing with people from other states with modern knowledge.
- Increasing migration from rural areas: The lack of opportunities in the state in jobs and education forces people to move out of the state. As more and more people are moving out from rural areas to urban places, a part of culture, tradition, and social customs are getting lost in the way. Increasing westernization of society and culture infusion has also demotivated people to follow their culture.

4.4 State of Financial Inclusion in the Rural Himachal Pradesh

Financial inclusion and social inclusion are essential parameters for determining the status of human development. These parameters also indicate the effectiveness of the policies of the government. The study aims to use rural tourism as a means to

Table 4.3 Status of financial inclusion in Himachal Pradesh

S. No	District	C-D ratio	Average population per bank	Number of banks
1.	Bilaspur	25.28	4602	83
2.	Chamba	26.54	6107	85
3.	Hamirpur	18.26	3728	122
4.	Kangra	22.33	5171	292
5.	Kinnaur	39.09	2629	32
6.	Kullu	38.63	4423	99
7.	Lahaul and Spiti	18.32	1973	16
8.	Mandi	24.71	5375	186
9.	Shimla	26.6	3119	261
10.	Sirmaur	69.94	5095	104
11.	Solan	61.45	2763	210
12.	Una	26.24	4493	116
	Total	30.8	4274	1606

Source: Author's compilation
Note: C-D means cash to deposit ratio

address the concerns of rural HP. It is essential that we first look at the present status of financial inclusion and social inclusion in rural Himachal Pradesh (Table 4.3).

4.4.1 Status of Financial Inclusion in HP

If one talks about the average population per branch, then a branch in Himachal covers an average population of 3202. However, the same statistic for India is approx. 11,000 as per RBI. This indicates that the ratio is significantly low in HP.

At present, 1848 Bank Mitras are deployed by various banks to provide essential banking services in villages. Banks have 10.55 lakh basic saving bank deposit accounts (BSBDA) under the scheme up to September 2017. Out of 10.55 lakh PMJDY accounts, banks have opened 9.20 lakh accounts in rural areas. In the state, banks have issued RuPay Debit Cards to 8.06 lakh PMJDY account holders and thus covered more than 78 percent of PMJDY accounts. Banks have taken the initiative to link the bank account with Aadhaar number and mobile number and seeded 88 percent of PMJDY accounts up to September 2017.

Financial literacy and awareness campaign plays a significant role in reaching the target groups. Banks are conducting financial literacy campaign through the financial literacy centers (FLCs) and through its bank branches in Himachal Pradesh. There are 22 financial literacy centers (FLCs) managed by Leads Banks, i.e., PNB/SBI/UCO and cooperative sector banks, viz., HP State Coop Bank (HPSCB), Jogindra Central Coop Bank (JCCB), and Kangra Central Coop Bank (KCCB). RBI has directed Rural Bank branches to regularly conduct the FLC camps at least once a month, focusing on banks and FLC's digital literacy and progress.

Table 4.4 Status of ATMs in Himachal Pradesh

Public Bank ATM	1443
Private Bank ATM	362
White Label ATM	44
Small Finance Bank ATM	6
Foreign Bank ATM	0
Payment Banks ATM	0
Total number of ATM in HP	**1855**

Source: Author's compilation

Automatic teller machines improve the accessibility of financial services to the general public. They ensure 24*7 availability of several services that the branches cannot do. HP has a total of 1855 ATMs. Out of these, public banks have the majority while there is no ATM of foreign and payment banks (Table 4.4).

An ideal credit deposit ratio for the commercial bank is 80–90%. But in HP, the ratio stands at 30.9%. It indicates that people are not borrowing from the organized banking sector. It could mean two things: either people in HP do not require funds and therefore are not borrowing or they are borrowing from an unorganized sector. While the first scenario is favorable, it is not practically possible.

Consequently, it means the presence of an unorganized sector is more than the organized lending system. The average cash deposit ratio for scheduled commercial banks of the country was 76.9% at the same time. It clearly shows that banks in HP are lending far less than the banks in other states. Hence, the credit generation is also less as compared to the other states.

4.4.2 Use of Technology in Rural Himachal Pradesh

Technology has revolutionized our day-to-day activities. Technology makes operation effective and efficient. Be it manufacturing or a service industry, technology is used.

4.4.3 Status of Electrification in Rural HP

For technology to exist in a place, electricity is a must. The absence of electricity acts as a hurdle in adopting technology. If we look at electrification status, it shows that there is complete electrification in rural areas. However, the number of households actually using electricity might differ. This is because people might not indulge in its consumption as they do not have the resources. The statistics for rural electrification are as follows:

- Total households in the state: 14,86,340
- Households electrified till 10 October 2017: 14,73,534

- Subsequent progress till date: 12,806
- Total achievement: 12,806
- Balance un-electrified households: nil

4.4.4 Use of Technology in Agriculture

Himachal Pradesh is a dominant agricultural state. The present agribusiness regularly utilizes complex advancements, for example, robots, temperature and dampness sensors, airborne pictures, and GPS innovation. These smart devices and meticulousness farming and mechanical frameworks permit organizations to be more productive, proficient, more secure, and all the more naturally agreeable.

- Irrigation techniques: Changes and advances in the water system industry have prompted new innovations that bring dribble water system's advantages inside ranchers' compass everywhere. New water system frameworks which use pressure redressing (PC) drippers require less siphoning power, further bringing down energy expenses and making sun-based fueled water system a plausible arrangement.
- Farming techniques: Ranchers settle on choices dependent on the data they have close by, which is why information has helped them bridle the intensity of data to settle on better-educated choices that permit them to utilize assets more economically. Observing and controlling harvest water system frameworks using cell phones, ultrasounds for animals, usage of versatile innovation and cameras, and crop sensors are many ways ranchers are utilizing innovation to upgrade their profitability.
- Purchase and sale of crops in e-Mandi: With the concept of e-Mandi becoming popular in the country after the introduction of New Farm laws, farmers are bound to use e-mandi for the purchase of raw materials and selling their produce.
- Payment acceptance: Accepting and making payment is one of the ways farmers are using technology. They are getting used to the habit, and more opportunities will help build their self-confidence for technology use.

4.4.5 Use of Technology in Service Industry in HP

Another industry that has its presence in HP is the service industry. With the absence of the manufacturing sector, people are either self-occupied or resort to the primary sector. Hence, it is essential to know if technology is playing any role in this sector. Technology plays a role in the service sector in the following manner:

- Digital marketing: Digital marketing contains adaptable methodologies and arising advances that advance often. Showcasing can be improved through programming updates with flexible instalment plans. Site support and substance advertising for clients help in holding the clients.

- Accepting digital payments: Digital payments significantly impact economic growth in the short run (Ravikumar et al. 2019; Kumar et al. 2023). Several industries in the service industry rely primarily on the services they provide to the digital payment industry. In addition, digital payments have made it easier to maintain cash balances and accounts.
- Building competitive advantage: Technology helps lower costs, work on differentiation, and focus on specialization. Thus, it helps in building a sustainable competitive advantage for the service industry.
- Increasing product base: The associations that are best in commercializing innovation and creating items have cognizant inside innovation and item guides that connect market, innovation, things, and their multi-year activity plans. Model-based designing is expanding in each space and capacity, from framework engineering to creation, and for the two frameworks and programming. Those associations best intentionally build up a device and cycle reconciliation of framework designing, programming designing, item information, the executives, check and approval, discipline models, and survey measures.

4.5 Tourism Typologies Possible/Practiced in Rural HP

Tourism in rural areas is described as the tourist visit to a rural area for active participation in a rural lifestyle. It is described as a travel activity in the premise close to the rural region. It includes consuming nature-based activities such as hunting and fishing, agriculture produce, walking in the jungle, desert safari, living local lifestyle, adopting local customs and traditions, consuming regional cuisines, using local handicrafts, enjoying local festivals and folklores, and living and residing in local ambiance rather than in a hotel or a resort. It is traveling to get close to nature in rural premises. Various types of tourism can be practiced in rural Himachal Pradesh. These typologies are discussed in the following section of the study.

4.5.1 Agro-tourism

The most common and prominent tourism typology that is practiced and developed in rural areas is the agro-tourism. The tourist takes active participation in agriculture and non-agriculture activities while staying in or near the farm.

Source: http://www.shalviriverresort.com/agri-tourism.html

As agriculture is the most dominant employer in the HP, there are ample opportunities to develop the agro-tourism. Such tourism can be practiced in the apple orchards, for which the state is world famous. In addition, the agro-tourism marketing is more comfortable as it is the strength of the state that has to be marketed.

4.5.2 Wellness Tourism

It is defined as travel to promote health and well-being through physical, psychological, or spiritual activities. The objective of wellness tourism is to improve and maintain health and quality of life. Increasing the level of stress, health challenges, and other mental issues has increased wellness tourism demand.

Source: https://pxhere.com/en/photo/1436281

A significant requirement by the tourist for wellness tourism is the availability of a pristine, clean, and calm environment and abundant of resource in HP. In addition, staying away from technology is a crucial prerequisite in wellness tourism. With limited infrastructure available in rural HP, it is much easier to limit the use of technology in day-to-day activities. Further, by engaging in other rural activities, a tourist can shift the focus from a stressed lifestyle and live peacefully. Rural HP can provide such destinations free from chaos, are limitedly accessible for tourists, have the necessary infrastructure available, and provide mental solitude for tourists. A significant advantage for the tourist is that the cost of visiting such destinations would also be economical.

4.5.3 Leisure Tourism

One of the most popular types of tourism comprises nature tours and visit to cultural heritage and sites for recreation and enjoyment. People seldom visit theme parks and leisure programs in the search of stress-free environment from their hectic schedule. In the pandemic era, where most people were locked down in their homes, people became anxious.

Source: https://toib.b-cdn.net/wp-content/uploads/2018/01/shimla.jpg

Leisure tourism is an activity that can be practiced throughout Himachal Pradesh without any compromise in satisfaction. As destinations are available in each state district, a tourist has various destinations to choose from. While leisure tourism is the most consumed and promoted tourism typology in HP, the policymakers should take care of the exhausting destinations. Any incident that can reduce tourist satisfaction must be restricted. For this, a check on the resource availability is necessary.

Further, underexplored destinations should be marked and promoted for tourism in rural HP. The economic gains from leisure tourism are more than the other tourism typologies, so it will be a key to develop the rural areas of the state economically. Some additional infrastructure is required to ensure its development. So, the infrastructure development should be carefully monitored to avoid any significant damage to the environment or the tourist satisfaction level.

4.5.4 Spiritual Tourism

Spiritual tourism refers to travel to find purpose and meaning in life. It elevates your physical, mental, and emotional energies. It develops, maintains, and improves your body, mind, and spirit. Briefly, it connects your body, mind, and soul. Hence, spiritual tourism can be defined as traveling on pilgrimage to acquire enlightenment, entertainment, and education to leave your footprints and take your memories back to your home by expressing your gratitude to God. Spiritual tourism includes the following spirituality in daily routine, consuming sattvic food, and changing daily habits. Spiritual tourism aims to live in an ambiance of purity, which leads to spiritual transformation.

Source: https://www.tourmyindia.com/images/monasteries-in-himachal.jpg

Himachal Pradesh is famous for spiritual attainment. With wellness tourism being a significant hit, destinations are suitable for the conducive development of spiritual tourism. The increasing stress level and anxiety orders have further boosted the demand for spiritual tourism. The present crisis for humanity has also given people time to introspect their life. This has further increased the chances of revival of spiritual tourism in the state. The factors prompting tourists to undertake spiritual tourism are favorably positioned in the state. Hence, spiritual tourism can be developed with ease in rural areas.

4.5.5 Adventure Tourism

A significantly popular form of tourism among youth worldwide, adventure tourism has its share in attracting tourists. To experience the adrenaline rushing through their veins, youth want to engage in activities that provide fun, joy, and excitement. It includes river rafting, paragliding, trekking and hiking, etc.

Source: http://hptdc.in/wp-content/uploads/2016/11/skiing-1.jpg

Adventure tourism is an activity undertaken by the youth. There is an urgent need to develop more avenues for adventure tourism to attract youth to the state. Adventure tourism will also help in achieving the objective of destination discontinuity. Adventure tourists are seeking new avenues and destinations for tourism. By developing alternate tourism destinations, tourist traffic from exhausting destinations can be shifted to underexplored regions without marketing efforts. It will help the tourism industry's sustainable and holistic development, essential for the modern-day tourism industry.

4.5.6 Religious Tourism

Tours and trips for meeting religious purposes are among the most consumed tourism typologies, particularly in developing and underdeveloped countries. People across the globe move to cities, countries, or even continents for religious tourism.

Source: https://static.india.com/wp-content/uploads/2014/09/Capture-25.jpg

Religious tourism offers community participation. It is a place where there can be a diffusion of knowledge, culture, and values from one person to another. It also helps people to acclimatize with the culture of a place. Community participation can also be used in the development of community capacity building. Religious tourism can be quickly developed in rural areas of the state. Temples located in underexplored regions can be marketed, and tourists can be attracted to those places. There are various places of religious importance situated all across the state. Thus, there is no shortage of supply side for the development of religious tourism in rural areas.

Further, rural tourism is practiced by the locals as well. Thus, maintaining infrastructure will not be a challenge for policymakers. The only challenge will be to ensure that the infrastructure's quality does not deplete with the locals' usage. With people of various religions living in the state, some destinations can suit various tourists' needs. The significant point is that it should be kept in mind that the focus should be on building destinations at places in rural areas where the current flow of tourists is less. Moreover, if tourists are welcomed, they will be served without losing their satisfaction level.

One of the subsets of religious tourism is pilgrimage tourism. In this type, people visit places that have the utmost importance in their religion. The significant difference between a pilgrimage tourist and a religious tourist is the extent of faith. Pilgrimage tourism is consumed at the maximum level of faith of a devoted individual.

Source: https://ihplb.b-cdn.net/wp-content/uploads/2014/09/Hidimba-Devi-temple.jpg

Religious and pilgrimage tourism can help tourists understand the state's culture, increasing social inclusion. Participation of women in tourism activities will further boost the women empowerment by making them self-reliant. The heads of strict associations have at first dismissed the intersection between tourism and religion. Simultaneously, they didn't confess to the way that explorers were viewed as travelers according to an economic viewpoint.

4.5.7 Wedding Tourism

It is defined as traveling outside the bride and groom's residence for a wedding. Another term used more commonly for the same is the destination wedding. Wedding tourism is on a roll because of a rise in the middle class's disposable income and destination wedding promotion in movies and TV shows. Owing to the emergence of wedding planners and supportive infrastructure, wedding tourism is one of the top niche tourism products in the country.

Source: https://media.weddingz.in/images/46cdcca2823caf09959cca600c762a9e/budget-destination-wedding-in-himachal.jpg

Rural regions in HP flourish with geographical and scenic beauty. In addition, they have farmhouses that can support the tourist demand for destination weddings. Since the destinations in HP are economical over other states, the popularity of HP for wedding tourism is also increasing. For wedding tourism to flourish, promotion of the industry should be done through endorsement in movies. Movies and cinemas inspire people to select a location for their wedding. In addition, such a typology of tourism is a particular category for high-income group tourists. The promotion easily influences tourists' segment through movies, and, hence, demand can be easily created. High-quality boarding lodging and food facilities are required at the destination wedding. Such an infrastructure is not usually found in rural areas. However, if developed strategically, then these destinations can flourish in the long run as well. These tourists will also be economically superior to most other tourists' segmentation, and expenditure will be more. It will result in economic gains for the industry. In addition, by witnessing people's lifestyles, the locals can adapt certain things. This will help in cultural diffusion and upliftment of the standard of living of people.

4.5.8 Eco-tourism

Eco-tourism is now defined as "responsible travel to natural areas that conserves the environment, sustains the well-being of the local people, and involves interpretation and education" (TIES 2015). Eco-tourism is a form of tourism involving visiting fragile, pristine, and relatively undisturbed natural areas, intended as a low-impact and often small-scale alternative to standard commercial mass tourism.

Source: https://thenewshimachal.com/wp-content/uploads/2018/06/eco-tourism-site.jpg

Eco-tourism is an umbrella term for all the tourism activities undertaken to engage with nature in environmental premises. The key difference between rural tourism and eco-tourism is that tourism needs to be called eco-tourism. It is essential to be undertaken in proximity to the environment. This environment may be in a rural or urban setting as well. However, rural tourism is tourism that is practiced only in rural regions. Himachal Pradesh has one of the cleanest environments in the country. It has clean air, provides clean drinking water, and is therefore best suited for eco-tourism development. The environment supports the incoming tourists and rejuvenates them by providing mental peace and relief from other diseases. The culture and values also help people respect toward Mother Nature and understand the valuable contribution of nature to human civilization.

4.5.9 Food Tourism

It is also known as gastronomic or culinary tourism. It is defined as tourism for the sole purpose of exploring the food items. It is about what is unique, authentic, and memorable about the delicious stories a destination has to tell. It includes farmers, cheesemongers, fishers, brewers, winemakers, and everyone in between. The top food tourism activities include dining out at a government restaurant, eating street and local food, and visiting a famous restaurant and bar.

Source: https://www.travelogyindia.com/storage/app/upload/traditional-himachal-foods.png

Rural HP has numerous stories to tell when it comes to food. Himachali Dham, a simple dish, is cooked in every district in its recipe. Further, the state boasts of an Indo-Tibetan cuisine famous for many Buddhist tourists visiting the state. Several dishes are popular in the state that suits the tourists with contrasting taste buds. Rural regions can act as the epicenter for food tourism development. It is mainly due to fresh, organic food items like vegetables and fruits, availability of authentic traditional spices, and knowledge of conventional food cooking techniques. Food tourism or gastronomic tourism is yet to be popularized in India. If HP can take an advantage, it can persuade daily visitors or young people to spend more. Food tourism also offers knowledge about the food items and culture as well. Eating together also increases bonding with the people. Thus, when tourists eat with local people in their backyard, it will increase adhesion and enhance social inclusion.

4.6 Environmental Challenges

Given the enormous advantages of rural travel tourism in job creation and social and economic development, de facto environmental conservation needs to be fostered as a devoted plan for comprehensive developmental plans. The significant issues and difficulties of rural tourism industry are the need to save the environmental assets, the need of training, legitimate comprehension for the all the stakeholders, and the need to create a development wherein the individuals at all levels are required to take part in the advancement of tourism.

- Environmental assets: The optimum utilization of ecological assets and services should be given high priority. Keeping in mind the per capita carbon footprints, the sustainable consumption patterns must fall in place.
- Over-tourism: Over-tourism can potentially cause enormous damage to the growth of the tourism sector. Pro-environment culture may likewise be utilized as an essential component of the travel industry.
- Suggestive framework: The long-term environmental consequences can pose a big threat, and hence mitigation plans or other strategies must be inclusive of the agenda/framework and other designing of the policies.

4.7 Conclusion

Notwithstanding arising research on clever portability arrangements in metropolitan regions, there have been not many endeavors to investigate the importance and manageability of these arrangements in rural settings. Besides, existing examination addressing rustic versatility arrangements normally centers around a specific client bunch, like nearby inhabitants, second-mortgage holders, or sightseers. In this paper, we concentrate on the social inclusivity, financial reasonability, and ecological effects of novel mobility arrangements in rural exploration. The study concludes that sustainable rural tourism can be effectively developed in rural Himachal Pradesh. This is because the state has the necessary human power available at its disposal. There are approximately an equal number of female populations in the state. This negates the concerns of female safety and failure to serve female tourists in the rural areas. The literacy rate is also sufficiently high in rural areas. It means both men and women are educated and can understand training methods to serve tourists. It also means that they can use technology if they are trained for it. The sex ratio in the state is too high. It means that the state has a favorable outlook for the females in the rural areas. Thus, female safety will not be a challenge for policymakers. The use of technology in the state is also adequate. The agriculture sector and service sector are using the technology in their present avenues. It means that introduction of technology to rural people will not be a significant challenge. Financial inclusion in the state is at an average level. It can be increased further by giving more opportunities to the rural people. Rural tourism can be an alternate to the existing schemes and techniques for promoting financial inclusion. The education level in rural HP indicates that people can transmit the social values to the incoming tourists. Moreover, they can understand good and bad values and can adapt them to increase their living standards. Thus, rural tourism should be promoted in the state.

References

Kumar, S., Kumar, V., Kumari Bhatt, I., Kumar, S., & Attri, K. (2023). Digital transformation in tourism sector: trends and future perspectives from a bibliometric-content analysis. *Journal of Hospitality and Tourism Insights*, Vol. ahead-of-print No. ahead-of-print. https://doi.org/10.1108/JHTI-10-2022-0472

Ravikumar, T., Suresha, B., Sriram, M., & Rajesh, R. (2019). Impact of Digital Payments on Economic Growth: Evidence from India. *International Journal of Innovative Technology and Exploring Engineering* 8(12): 553–557.

Understanding the Perception of Tourist on the Development of Rural Tourism in Himachal Pradesh

5

5.1 Introduction

For a product to be successful, it must be demanded by the tourists. A product will fail to achieve the target if it does not meet the expectations of potential customers. Tourists visiting Himachal Pradesh will be the potential customers for rural tourism product (Shrivastava and Heredge 2004; Bramwell 1994). Therefore, it is imperative that an understanding of their perception and behavior exhibited by them is undertaken to be incorporated while developing the tourism product. Tourism industry offers both opportunities and challenges for gender equality and other environmental factors in achieving the sustainable development (Cole and Sinclair 2002; Bird and Sapp 2004). In addition, it is essential to understand the typology of tourists that visit the state, since tourists' demographic factors such as age group, income group, the preferred season of visit, and educational qualification have impacted the tourist behavior (Kumar 1993; Dreze and Sen 2002). Subsequently, this chapter will present the relationship that these factors exhibit with the behavioral aspect such as estimated budget, willingness to deviate from the budget, the average length of stay, destination visited, and type of tourism typology availed. Further, per trip's economic contribution is derived to determine the possible earnings for rural tourism and a chance for them to make themselves competitive in terms of quality and pricing. The following table records the demographic attributes of the tourists surveyed for the study (Table 5.1).

The table features that numerous tourists visiting Himachal Pradesh belong to the age group of 25–35 years. This age group is either well settled in their job, has started their married life, or has the growth stage in their working life. This is likewise clarified through the tourists' income group as the central part of the respondents belonged to middle- or higher-income groups, i.e., from 3 lakh to 8 lakh and more than 8 lakh income per annum. This is beneficial for the state that high-earning people also visit. However, the data for the low-income group is not encouraging. This means the tourism products offered by the state are not pretty much inclusive

S. Kumar, *Sustainable Rural Tourism in Himalayan Foothills*,
https://doi.org/10.1007/978-3-031-40098-8_5

Table 5.1 Descriptive statistics of respondents (tourists)

S. No	Demographic factors/key statistics	Variables and percentage
1.	Age group	Below 18 years (1%)
		18–25 years (18%)
		25–35 years (52%)
		35–45 years (20%)
		More than 45 years (9%)
2.	Gender	Male (63%)
		Female (37%)
3.	Educational qualifications	Graduation (1%)
		Post-graduation (1%)
		Professional courses (25%) Secondary education (61%)
		Senior secondary education (14%)
4.	Income group	1 lakh–1.5 lakh (7%)
		1.5 lakh–3 lakh (16%)
		3 lakh–8 lakh (31%)
		More than 8 lakh (46%)
5.	Nationality	National (75%)
		International (25%)
6.	Preferred season of visit	January to march (20%)
		April to June (41%)
		July to September (9%)
		October to December (31%)
7.	Average duration of stay	4.08 days
8.	Average expenditure made per trip	Rs33907
9.	Average tourist group size	4 persons

Source: Primary survey

in nature. There is a need to develop product offerings that cater to the lower-income group and be suitable for the high-income group. The preferred season to visit the state is April to June, as people visit the state to relax in the cool hill stations. The period of October to December is also among the top choices. People like to spend their Christmas and new year celebrating in the snow-laden mountains. A key finding from the survey was that the average duration of the tourists' stay in the state was about 4 days. This statistic is more prominent as this is greater than the national average duration of tourist stay. In addition, it is statistically found that the more days a tourist spends at a destination, the higher are the earnings, keeping in mind sustainability concerns. It is also tracked down that the average tourist group size is four people and people like to visit the state with friends or families. Solo traveling is not popular among any age group; however, it has an extreme potential for development. The average expenditure per trip per group (including boarding, lodging, food, shopping, and others) was found to be Rs 33,907. The stakeholders' aim should now be to enhance the expenditure made by the tourists, in case there is a need to restrict the number of tourists while simultaneously maintaining the price competitiveness of the industry to other adjoining states. The following sections will establish the relationship between the tourists' demographic aspects with their behavioral elements to get into useful insights for the industry.

5.2 Data Interpretation

The primary data collected with the help of structured questionnaire from the tourists who visited the state of Himachal Pradesh has been analyzed in the following sub-headings in the given paragraphs.

The data collection done from the tourists across both genders with their education level is presented below.

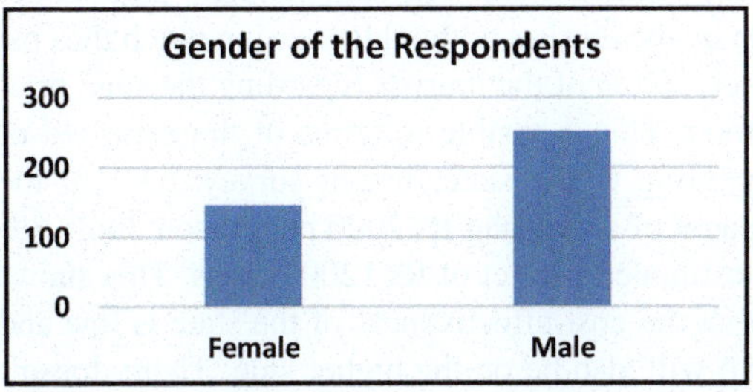

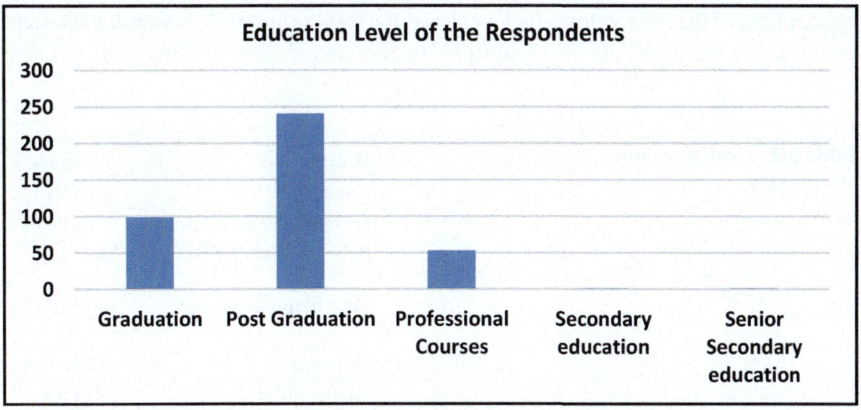

5.2.1 Place of Stay

On being asked where they were residing during their journey, most respondents chose private hotels as their answer. This is because the private hotels are more comfortable to search and get reviews through social media. Their booking is easy and hassle-free. They also have tie-ups with the travel agencies and transporters through which people get their tour planned. Homestays are also preferred by the people for their stays. This highlights that the government scheme of developing homestay is yielding positive results for the industry. Government hotels, guest-houses, and circuit houses are also chosen by the people for their stay because of low tariffs, quality, and security assurance (Table 5.2).

5.2.2 Budget of Tourists

The tourism industry survives on the tourists' economic contribution in the form of boarding, lodging, shopping, and food expenditure. The stakeholders' aim should be to enhance the expenditure made by the tourists without increasing much the cost of the itineraries. In addition, the perception of the tourists regarding the state pricing mechanism should be understood. This is visible in terms of the expected or estimated budget of the tourists coming to the state. In our survey, 62% of the respondents had their estimated budget of more than Rs 2000 per person. Only 8% of the respondents had per person estimated budget of Rs 1200 or less. This shows that tourists visiting the state are that the cost-effectiveness of the state is low and expenditure incurred during the trip will also be on the higher side. There doesn't appear to be much significant difference in actual expenditure made. Only 9% of the tourists had a substantial expenditure of Rs 1200 or less during their stay, while greater part of the respondents had an actual Rs 1600 or more. A reason for the same could be the increased average length of the remain calculated (Table 5.3).

Table 5.2 Tourists preferred place to stay

Place of stay	Frequency
Campsite	20 (5%)
Government hotels/guest houses/circuit house	75 (19%)
Own home/relatives	14 (4%)
Homestay	95 (24%)
Paying guest	2 (1%)
Private hotel	194 (49%)
Total	**400**

Source: Primary survey

Table 5.3 Expenditure incurred by the tourists

S. No	Budget range	Estimated expenditure	Actual expenditure
1.	400–800	11 (3%)	19 (5%)
2.	800–1200	21 (5%)	15 (4%)
3.	1200–1600	64 (16%)	48 (12%)
4.	1600–2000	58 (15%)	90 (23%)
5.	More than 2000	246 (62%)	228 (57%)
	Total	**400**	**400**

Source: Primary survey

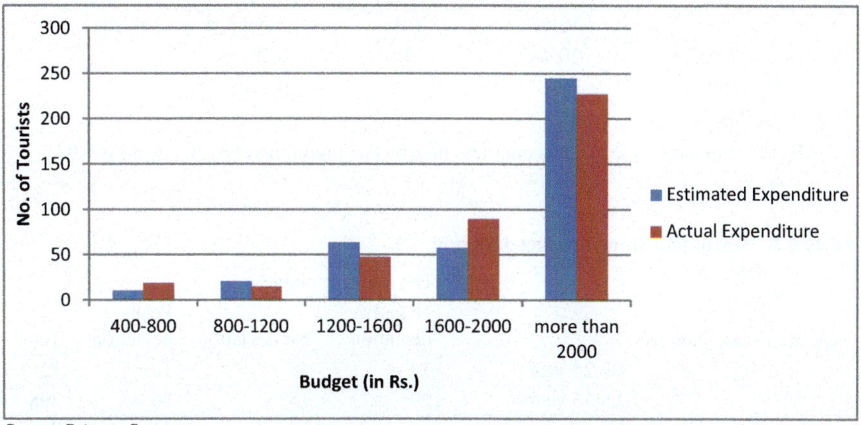

Source: Primary Survey

Fig. 5.1 Expenditure by the tourists. (Source: Primary survey)

5.3 Comparison of Estimated and Actual Expenditure by Tourists

The graph below compares the estimated and the actual expenditure incurred by the tourists during their stay. Only the respondents with an estimated budget of Rs 1600–2000 had significant variation. Only 15% estimated their expenditure to be within the bracket (Fig. 5.1).

Since the stakeholder's concern is with the tourists' amount during their stay, it is essential to understand which group of tourists spend more during their visit. For analysis, we assume the following hypothesis (Table 5.4).

Ho1: *The expenditure made by the tourists is independent of the age group of the tourists*

On running the chi-square test for independence between the age group of tourists and the actual expenditure made by them, the calculated chi-square value at significance value 0.05 is found to be 0.210. As the calculated statistic is more than the significance value, the null hypothesis is accepted. This means that the industry should focus on each of the tourists' age groups without no attention on a specific section.

Table 5.4 Age group of tourists and expenditure made

Age of the respondent	Mean (actual expenditure)					
	600	1000	1400	1800	2200	Total
18–25 years	5	5	10	17	35	**72**
25–35 years	7	6	30	42	123	**208**
35–45 years	5	2	5	26	43	**81**
Below 18 years	0	0	1	0	1	**2**
More than 45 years	2	2	2	5	26	**37**
Total	**19**	**15**	**48**	**90**	**228**	**400**

Chi-square tests

	Value	df	Asymp. Sig. (2-sided)
Pearson chi-square	20.449[a]	16	0.201
Likelihood ratio	20.222	16	0.210
N of valid cases	400		

[a]12 cells (48.0%) have an expected count less than 5. The minimum expected count is 0.08

Table 5.5 Age of tourists and budget deviation

Age of the respondents		Deviation in budget			
		Negative deviation	No deviation	Positive deviation	Total
	18–25 years	12	48	12	**72**
	25–35 years	29	153	26	**208**
	35–45 years	15	55	11	**81**
	More than 45 years	3	30	4	**37**
	Below 18 years	1	1	0	**2**
Total		**60**	**287**	**53**	**400**

Chi-square tests

	Value	df	Asymp. Sig. (2-sided)
Pearson chi-square	6.025[a]	8	0.644
Likelihood ratio	5.824	8	0.667
N of valid cases	400		

Source: Primary survey
[a]4 cells (26.7%) have expected count less than 5. The minimum expected count is 0.27

Apart from the actual expenditure made, a vital aspect of the tourist is their willingness to deviate from their budget. Suppose the tourists accept deviation in the budget and also are happy to increase their spending. In that case, the stakeholders could lure the tourists into spending more by offering more products or unique products. It is also crucial to know if a particular age group is more acceptable to accept the deviation in the budget than others. For this, a chi-square test is conducted to know if there is any relationship between tourists' age and their willingness to accept budget deviation (Table 5.5).

Ho2: *The desire to accept budget deviation is independent of the age of the tourists*

From the table, it is seen that around 72% of the tourists faced no deviation in their budget. Their actual expenditure was within their estimated expenditure. Fifteen percent of tourists had estimated more than what they spent, and approximately 13% of tourists spend more than they estimated. This shows that the industry is not fully able to lure tourists into spending more. It is common that when we go shopping, we tend to go over budget for the products of our liking. In this way, the industry lost much business by not offering products and services that could generate more economic contribution. On running the chi-square test, the calculated chi-square value at a significance level of 0.05 was found to be 0.644. As the calculated value is more than the significance value, we have to accept the null hypothesis. This means that each age group is willing to accept the deviation from their estimated budget.

5.3.1 Frequency of Visit to the State

A destination is thriving only on the off chance that it can attract tourists again and again. Inability to do so means that the tourists' expectation was not fulfilled and there is a need to work upon the offering in terms of product or pricing. More than 64% of the tourists have visited the state in our survey more than five times. This implies that tourists like the products offered and therefore visit the destinations time and again.

It is also essential to understand if the willingness to visit a destination time and again depends on the tourists' age group. For this, a chi-square test has been conducted to determine the influence of the tourists' age on the recurrence of tourist visits (Table 5.6).

Table 5.6 Age group and frequency of visit of tourists

		How many times have you visited Himachal Pradesh as a tourist?				
Age group		First visit	First to fourth visit	Fifth to eighth visit	More than eighth visit	Total
	Below 18 years	1	0	1	0	**2**
	18–25 years	10	33	6	23	**72**
	25–35 years	6	60	44	98	**208**
	35–45 years	4	17	11	49	**81**
	More than 45 years	1	12	10	14	**37**
Total		**22**	**122**	**72**	**184**	**400**

Chi-square tests

	Value	df	Asymp. Sig. (2-sided)
Pearson chi-square	45.251[a]	12	0.000
Likelihood ratio	40.106	12	0.000
N of valid cases	400		

Source: Primary survey

[a]7 cells (35.0%) have expected count less than 5. The minimum expected count is 0.11

Ho3: *Frequency of visit to Himachal Pradesh is independent of the age of the tourist*

On running the chi-square test of independence, the calculated chi-square value at a significance level of 0.05 was found to be 0.00, which is less than the significance value. Therefore, the null hypothesis is rejected. This means that a particular age group of people visit HP more frequently than the other age group.

It is also beneficial for the industry if the often-visiting tourists belong to the high-income group. People with high income can spend more than those with low income. Therefore, it is also essential to know that if a particular income group tourist visits the destination frequently, they find something consistent in the product offering. For this purpose, a chi-square test of independence is conducted to determine the relationship between tourists' income group and their frequency of visitors to the state (Table 5.7).

Ho4: *The frequency of visits to the state is independent of the income group of the tourist*

From the table, it is seen that people belonging to the higher-income group (3 lakh to 8 lakh and more than 8 lakh) are the ones that are repetitive visitors to the state. It is a good sign for the industry that people with high disposable income are more frequent to the state. On running the chi-square test, the statistic's calculated value is found to be 0.08, which is less than the significance value of 0.05. Therefore, the null hypothesis is rejected. This means that there exists a significant difference between income groups concerning the frequency of visits.

Table 5.7 Income of tourists and frequency of visit of tourists

		How many times have you visited Himachal Pradesh as a tourist?				
Annual family income		First visit	Second to fourth visit	Fifth to eighth visit	More than eighth visit	Total
	1 lakh to 1.5 lakh	2	10	7	12	**31**
	1.5 lakh to 3 lakh	4	25	12	21	**62**
	3 lakh to 8 lakh	10	41	9	62	**122**
	More than 8 lakh	6	46	44	89	**185**
Total		**22**	**122**	**72**	**184**	**400**

Chi-square tests

	Value	df	Asymp. Sig. (2-sided)
Pearson chi-square	22.431[a]	9	0.008
Likelihood ratio	24.512	9	0.004
N of valid cases	400		

Source: Primary survey

[a]2 cells (12.5%) have expected count less than 5. The minimum expected count is 1.71

5.3.2 Duration of Stay

Duration of stay is a significant aspect of the tourism industry. How long a tourist stays has a crucial impact on the economic contribution made by the tourists. Longer durations of tourists mean more earnings even though the number of tourists reduces (Ngoc et al. 2023). The primary concern for the industry is on finding ways to increase the length of tourists. In India, it is estimated that the average length of tourist stay for domestic tourist is 2.6 days and for foreign tourists is around 3.2 days. In our study, the average stay duration was 4.08 days, which is higher than the national average. This means that tourists like to spend more time in the state because of multiple reasons like price competitiveness, diverse product offering, and a cool environment. However, enhancing the length of stay will cause a restriction on the number of tourists that the industry could serve. This will prompt the inactive limit of a portion of the assets. Therefore, it is crucial to strike out the balance between the average length of stay of tourists and the number of catered tourists. The following table lists the various durations of stay that the respondents had during their trip (Table 5.8).

To expand the duration of stay, it is essential to know if the seasonality of visits influences tourists' stay. Assume a tourist stays longer in a particular season; in that case, the reasons behind so must be identified, and corrective actions should be taken for other seasons as well. For the purpose mentioned above, a chi-square test has been conducted to determine the reliance of the average duration of tourist stay on the seasonality of visit (Table 5.9).

Ho5: *The average duration of tourist stay is independent of the season of visit*

A null hypothesis is tested to determine the relationship between the average duration of stay and the visit seasonality. The calculated chi-square statistic is found to be 0.001, which is lower than the significance value of 0.05. Therefore, the null hypothesis is rejected. This means that a significant relationship between the seasonality of visit and the average duration of the stay exists.

Table 5.8 Duration of tourist stay

S. No	Duration of stay	Frequency
1.	Overnight stay	19
2.	2–3 days	174
3.	4–5 days	117
4.	5–7 days	52
5.	More than a week	38
	Grand total	400

Source: Primary survey

5.3.3 Average Person Per Trip

Another significant aspect of tourists is the size of the tourist group. Our survey had five tourist groups, namely, solo traveler, a group of 2–4 persons, 4–6 persons, 6–8 persons, and more than 8 persons. An average person per group was determined through exponential smoothing, and cross-tabulation was done (Table 5.10).

Ho6: *The tourist group size is independent of the season of the visit*

It is also essential to know if the average person on a trip is dependent on the season of visit. This shall provide the useful insight into the seasons that attract

Table 5.9 Season of visit and average duration of tourist stay

Season of visit		Average duration of stay in days					Total
		0.5	2.5	4.5	6.5	8.5	
	April–June	6	71	48	21	17	163
	January–March	1	30	26	8	16	81
	July–September	0	20	11	2	1	34
	October–December	12	53	32	21	4	122
Total		19	174	117	52	38	400

Chi-square tests			
	Value	df	Asymp. Sig. (2-sided)
Pearson chi-square	33.313[a]	12	0.001
Likelihood ratio	34.495	12	0.001
N of valid cases	400		

Source: Primary survey
[a]4 cells (20.0%) have an expected count less than 5. The minimum expected count is 1.62

Table 5.10 Season of visit and average person in a trip

Season of visit		Average person accompanied					Total
		1.0	3.0	5.0	7.0	9.0	
	April–June	15	122	17	2	7	**163**
	January–March	10	39	16	8	8	**81**
	July–September	8	20	2	1	3	**34**
	October–December	12	73	13	7	17	**122**
Total		**45**	**255**	**48**	**18**	**34**	**400**

Chi-square tests			
	Value	df	Asymp. Sig. (2-sided)
Pearson chi-square	31.578[a]	12	0.001
Likelihood ratio	31.484	12	0.003
N of valid cases	400		

Source: Primary survey
[a]3 cells (18.8%) have expected count less than 5. The minimum expected count is 3.83

more tourist size groups than individual tourists. People traveling in groups are more attractive than individual tourists as the potential income from a group is much more than an individual. A chi-square test with the following hypothesis was conducted to determine the relationship between seasonality of visit and tourist group size. On performing the chi-square test, the test statistic was calculated to be 0.001, which is lower than the significance value of 0.05. This means that the null hypothesis stands rejected. There is a significant relationship between the seasonality of visits and tourist group size.

5.4 Pre-visit Analysis

In this segment of the analysis, the various factors and indicators which affect the tourists' decision-making process regarding their visit to the particular destination has been discussed and analyzed.

5.4.1 Purpose of Visit

Himachal Pradesh offers diverse products in its tourism portfolio. Yet, simple contribution doesn't imply that industry is diversified in nature. For that, each offering must be consumed by the tourists. In our survey, the respondents were asked about the purpose of their current visit. They are likewise inquired as to why their past visit to know about the consumption pattern of typologies of tourism. This investigation additionally helped in determining the status of various tourism typologies in the state. There are chances that particular typologies, despite the fact that they are offered, are not devoured by tourists. In addition, it helps us in assessing which tourism typology is facing a lot of burden. Marketing strategies will be made, keeping in mind the consumption pattern of the tourists as well.

In our survey, leisure tourism, adventure tourism, spiritual tourism, education tourism, and religious tourism are the conspicuous tourism typologies consumed by the tourists. This is principally a result of different leisure tourism sites spread across the state because of its geographic characteristics. Adventure tourism is also popular in trekking in the mountains, camping in valleys, and rafting rivers. People also visit Himachal Pradesh for relaxing mentally and physically due to the stressful life of the cities. The presence of religious shrines also attract religious and spiritual tourism in the state. People also visit the state for educational purposes such as conducting research, drawing seminars, and studying university libraries. Wedding tourism is also a popular typology of tourism. Himachal Pradesh is a hit choice for destination weddings among the elite class. However, the respondents failed to identify or mention emerging niche tourism such as eco-tourism, rural tourism, dark tourism, wine tourism, food tourism, halal tourism, and others. This reveals to us that there is scope for its development. However, massive marketing efforts will be desirous to make people aware.

5.4.2 Planning of Trip

Planning of visit by the tourist offers useful marketing insights. It informs us about whether the various categories of people play a role in the decision to visit. The different types that could assist in planning trips are self-planning, planned by friends or relatives, prepared by elders in a family, planned by tour operators, or a spontaneous trip. Every group has different activities in their mind. Planning by older folks in a family will be more budget specific, covering destinations that suit each member. It will be more oriented toward religious and spiritual tourism. They tend to keep away the adventurous part of their tour planning. Self-planning indicates the emerging behavioral shift in the consumers. As youth become self-reliant and tech-savvy, they have started to explore things on their own and plan their itineraries. The unplanned or spontaneous trip is also an emerging phenomenon. Still, it is mostly practiced by youth who travel in small groups or prefer solo travel. Suppose a tour operator is hired for planning. In that case, there is a possibility that through tie-ups, the behavior of tourists can be altered (Table 5.11).

In our survey, self-planning emerged as the most favored mode of planning by the tourists. Technological advancement has made it quite possible for tourists to plan their travel hassle-free. This is also possible as the youth today try to be more creative, which requires flexibility. As youth has become more participative in decision-making, self-planning is the way to move ahead. Since most of the travel is done with friends, one friend, who either has more experience in traveling or has visited the destination before, is given the responsibility to plan the travel. Unplanned trips are common among the youth these days. This gives the flexibility of residing or choosing destinations based on the availability. This was also selected as a frequent choice in our survey. What is to be understood here is that since most of the planning for the tour is undertaken by self, tie-up with the social media platform to attract people is a must. In addition, informative content about your tourism product should be displayed across social media and over the Internet so that it is easier to access your content.

Table 5.11 Tour planning statistics

S. No	Planned by	Frequency
1.	Planned by yourself	185 (46%)
2.	Decided by friends	88 (22%)
3.	Unplanned trip	76 (19%)
4.	Decided by an elder in the family	23 (6%)
5.	Planned by a tour operator	11 (3%)
6.	Planned by relatives	17 (4%)
	Total	**400**

Source: Primary survey

Table 5.12 Planning period statistics

S. No	Planning period	Frequency
1.	A week ago	140 (35%)
2.	7–14 days	86 (22%)
3.	15–30 days	67 (17%)
4.	More than a month ago	107 (27%)
	Total	**400**

Source: Primary survey

5.4.3 Planning Is Done How Many Days before

Another aspect of tour planning is the number of days that are taken before a tour is finalized. Longer duration in taking decision means a meticulous analysis of budget and the interest of the tour members. Longer durations in planning are mostly visible when planning is done by the elders in the family or requires the consumption of various typologies in a limited budget available (Table 5.12).

There were either very short-term planners (planned only a week ago) or those who had planned more than a month ago in our survey. The plausible reason is that the people visiting the state have also previously visited the state before. People with a longer duration of planning usually have more people accompanying them so that the planning is done keeping in mind the interest of multitude of individuals in a group.

5.4.4 Perception of People About HP Tourism Industry (Suggestions by People Before Visiting)

It is equally important to know about the image that an enterprise has in its prospective customers' minds. This is significant since in such case, people are satisfied with the current offerings and a positive image for the overall industry is created. It is much easier to introduce new products under its portfolio and pass on to the customers. People are asked to provide suggestions that they received before visiting the destinations from their friends or family members. This is important as it tells us about the past experiences of tourists who had earlier visited or the increasing concerns faced by the tourists in the industry in general.

Our survey revealed that the significant suggestion received by the tourists is of the security threat. This indicates that people are hesitant to visit the state because of growing security concerns for the tourists throughout the country or in the state in particular. Several cases of theft and crimes against tourists have been reported in the state. In addition, tourists were advised not to go near the mountains and steep valleys in the dark, indicating a problem with road street lights in the state. Due to the concerns of turbulent cool weather, the tourists were suggested to keep woolen clothes with them if any such urgency arises. Other suggestions included taking photos and videos to keep memories, about tasting food, and enjoying particular destinations and other things. Overall, the reviews were positive about the state industry, which means that people usually enjoy their stay. This is the sole reason that we find a high frequency of revisits in our survey. This type of response is encouraging for the state tourism industry. This means that tourists have a favorable opinion of the industry. If a new tourism product is launched, tourists will be welcoming the product. They will not thwart in visiting new destinations.

5.4.5 Use of Social Media While Planning

Social media has emerged as an inevitable companion of tourists these days. It has empowered tourists by providing information that is sensitive enough to make a tourist change its decision. From posting about destinations to rave reviews about hotels and restaurant food items, the influencers these days can trap the mind of the potential tourists. Therefore, it is essential to understand the information that tourists search on social media before visiting a destination or consuming a tourism typology. Since the objective here is to provide information for the strategy formulation for rural tourism development, whatever tourists search over social media should also be known to the policymakers. By this, they can alter the minds of the upcoming tourists. They can lead them to places and tourism products they were unaware of or were reluctant to consume.

Our survey observed that the most frequent search on social media made by the tourist before visiting a destination is about the place to stay. It can be reading customer reviews about hotels, restaurants on personal blogs, recommendations by the influencers for a particular hotel or resort, or scrolling through various websites providing such information. The tourists also searched for information regarding the destinations such as famous places to visit, connectivity and accessibility of destinations, festivals and shopping places at a destination, and weather and climate conditions for the tour period. People also go through the videos and photos clicked by travel influencers to get the hotels and destinations' real-time experience. Thus, this tells us that many tourists engage in social media search before visiting a destination. So, suppose rural tourism is to be developed. In that case, it is to be ensured that proper social media management is undertaken, helping a tourist make rational decisions for their trip. Negative reviews must be countered and resolved so that the image of the tourism product is not diminished.

5.5 During Visit Analysis

Under this part of the study, the various factors have been analyzed which affect the tour of tourists (Table 5.13).

Table 5.13 During visit descriptive analysis of tourists' responses

Particulars	Frequency
Increase duration of stay	Yes ($n = 317$)
	No ($n = 83$)
Swapping current trip with new	Yes ($n = 301$)
	No ($n = 15$)
	Maybe ($n = 84$)
Services of a photographer	Yes ($n = 221$)
	No ($n = 179$)
Visited destinations on recommendation of locals	Yes ($n = 230$)
	No ($n = 170$)
Technology used	Yes ($n = 300$)
	No ($n = 100$)
Knowledge of "Dham"	Yes ($n = 273$)
	No ($n = 115$)
	Maybe ($n = 12$)
Food recommendation	Yes ($n = 305$)
	No ($n = 95$)
Consuming recommended food	Yes ($n = 275$)
	No ($n = 30$)
Services of food aggregators (Swiggy and Zomato)	Yes ($n = 57$)
	No ($n = 343$)
Mode of transportation	Use of cab aggregators ($n = 96$)
	Use of local transport ($n = 304$)
Experience with people	Extremely dissatisfied ($n = 4$)
	Moderately dissatisfied ($n = 0$)
	Slightly dissatisfied ($n = 7$)
	Neutral ($n = 11$)
	Slightly satisfied ($n = 60$)
	Moderately satisfied ($n = 136$)
	Extremely satisfied ($n = 182$)
Overall experience	Extremely dissatisfied ($n = 6$)
	Moderately dissatisfied ($n = 1$)
	Slightly dissatisfied ($n = 5$)
	Neutral ($n = 17$)
	Slightly satisfied ($n = 64$)
	Moderately satisfied ($n = 126$)
	Extremely satisfied ($n = 181$)

Source: Primary survey

5.5.1 Increasing Duration of Stay

As discussed, the duration of stay has its economic implications. A tourist, based on his budget, destination characteristics, time available, and his touring objectives, decides upon a length of stay. Our survey's average duration is more than the national estimate; however, efforts should be made to increase the stay. In addition, the willingness of a tourist to improve its stay is an important criterion. This means that the destination has provided more satisfaction than that was expected. In our survey, 79% of the tourist wished to increase their stay.

This response is encouraging for the tourism industry. It means that the industry is excelling in serving the tourist. The tourists' needs are fully satisfied, and they feel that they should spend more time and enjoy their stay. However, it also raises specific concerns. A tourist plans a visit well in advance after a careful analysis of social media and other networking platforms. Suppose a tourist finds something new that it did not encounter over the network. In that case, this means that the marketing of the industry is not done correctly. Therefore, there is a requirement for the marketing efforts to be strengthened and integrated. But this poses favorable opportunities for the rural tourism industry. As many of the respondents who are willing to increase their stay, they could be diverted to the rural destinations for determining if they liked the place and typology or not. The industry's willingness to increase should be captured by offering products that the tourist had never experienced before. Further, as rural tourism will be a less costly affair for the tourist, the tourist will be happy to stay for more without compromising much on the budget.

5.5.2 Places Visited During Stay

A tourist visits destinations that suit his objectives and are recommended by the family or friends with a high recommendation from tourist guides, travel agents, and travel influencers, or easily accessible. Himachal Pradesh boasts for several tourist destinations that are popular among the tourist from all across India. These destinations are present in several districts. However, there is an imbalance found in tourism across the districts of the state. Hence, it becomes essential to know about the destinations that are visited by the tourists and identify the overburdened destinations. Similarly, good things from these destinations like their attractiveness must be studied. Findings shall be implemented wherever possible at the other unexplored or underexplored destinations.

In our survey, the destinations that are frequently visited by the tourist include Shimla, Manali, Khajjiar, Dharamshala, Kullu, Kangra, Mcleodganj, Karsog valley, Rohtang pass, Rewalsar, Dharamshala, Kufri, Kasol, Indian Institute of Advanced Studies, Jwala Devi, Kamakhya Devi, Chamunda Devi, Parashar Lake, and Dalhousie. These destinations' critical aspect is that they either offer leisure tourism (sightseeing included), adventure tourism, or religious tourism. The results are consistent with the purpose of visit of tourists in our sample who arrived in Himachal Pradesh. Many tourists also visited educational institutes like IIAS in Shimla, a hub for conferences and seminars in the state. So MICE is also a popular niche tourism but is limited to micro-niches. These destinations are present in three districts of the state, i.e., Shimla, Kangra, and Kullu. This affirms the findings of other studies that the primary tourist footfall is in these three districts only. This means that these districts are overburdened with the tourists. So, there are chances that rural tourism development could further enhance the pressure on these industries. To develop rural tourism sustainably, the destinations must be identified in the districts with an abundance of natural resources, necessary infrastructure, and less tourist footfall. This will ensure the holistic development of the state tourism industry and assist the policymakers with accomplishing the goal of destination discontinuity.

5.5.3 Swapping Current Trip with New Unexplored Places in Himachal Pradesh?

For the effective implementation of destination discontinuity to a rural tourist destination, tourists' opinions must be assessed if they are willing to swap their current trip or not. Only if they wish to do so, then only the effective implementation of the desired objectives can be undertaken. In our survey, we inquired about the respondents if they wish to exchange their trip to visiting the rural tourism destination. The following responses are recorded for the question.

The responses are favorable for policymakers having the objective for the development of sustainable rural tourism development. The majority of people agreed to swap their current trips. People who rejected the suggestion were a minute 4% only. The others were not sure, but if effectively persuaded, they could have opted to accept the proposal. This means that the role of advertising and marketing will be much more in attracting tourists to rural areas than is thought of.

5.5.4 Destination Recommended and Visit

The primary reason for not visiting other destinations could be the non-marketing of such destinations. However, the primary concern arises when people are aware or are made aware of the destinations. Then also, they restrict themselves from visiting them. It is the role of stakeholders such as hoteliers, restaurant owners, tour guides, transportation sectors, and locals to make tourists aware of the unpopular tourist destination. Suppose the tourists decline to visit the tourist destinations recommended. In that case, that means that either there is a lack of trust between the parties or the destinations do not have the required infrastructure to attract the tourists. In our survey, the following responses were recorded to answer whether the tourist visited the destinations recommended by the other stakeholders or not.

The data shows some mixed results. Although most people visited the destinations recommended by the locals, the people adhering to their arrangements are likewise high. The probable reasons could be the lack of trust among the tourists and other stakeholders or the infrastructure's absence at the destination. It is also quite possible that the destinations recommended do not fulfill the objectives of the tourists. The ramifications for rural tourism development here are high. It clearly shows that although people were willing to visit rural destinations, they did not do it on their current trip. This means that to make rural tourism effective, it has to make a part of a package to search about it and avail it prior to beginning their visit. Abrupt change in their excursion, albeit wanted, however isn't valued and undertaken by the tourists visiting the state.

5.5.5 Use of Technology During Stay

Technology is one of the critical aspects of an individual's life today. It has shaped the way we behave and has created a paradigm shift in it. Tourists also make use of technology during their visit. It may be using GPS for navigation in unknown places, searching for weather and tourist destinations, using a language converter to communicate with the locals, ordering food items, making digital payments, capturing photos and videos on its smartphones, and booking cab services. The use of technology by tourists also depends on whether the other stakeholders are proficient with its use. If there is a balance between the two, technology can play a vital role in making the trip easy.

The data shows that 75% of the respondents used technology in a way or another during their stay. The reason for not using technology by others was the absence of mobile networks at certain places. The lack of an opportunity to use technology was also significant for not using technology at different locations. The results show that tourists in Himachal Pradesh prefer using technology. Still, if somehow they are unable to use it, it does not create dissatisfaction. In our study, it is observed that tourists have made use of technology for various purposes. Tourists believed that technology has made their journey more convenient.

Convenience in the booking of tour itineraries and services is one area where tourists made use of online services. The tourists were also able to check the best alternatives available for them through reading reviews on social blogs. Weather forecast, communication and photography, and information about the area were also the reasons tourists were using technology. One interesting observation was that people had used technology to search for public utilities like government toilets at tourist destinations. Again, they were using translators to communicate with shopkeepers in their language for better bargaining. Tourists also availed of discounts using online platforms such as Trivago and Goibibo. In short, the use of technology was made by tourists only for their convenience in booking and searching for information. Limited tourists made an online payment. This might be because the other stakeholders were not accepting the payments in digital mode. So while developing rural tourism, it should be noted that people using technology could be done manually, or tourists can be informed about the destination in brief through places. As rural areas comprise locals communicating in the native language, the availability of translators should be ensured.

5.5.6 Food Consumed During Stay

Local cuisines form an essential type of tourism typology. Food tourism is a growing niche tourism segment. For that, a dish should be selected that could be branded as a state dish and can be used to market the state's food tourism industry. Himachal Pradesh offers an assortment of food that differs from district to district. Each district has its local cuisine due to its cultural practices. It is of utmost significance that tourists consume the local dishes as well. Suppose the tourists consume food that is available throughout the country. In that case, this means that tourists have no knowledge about the local products or are hesitant to try them. The variety of Himachali food must reach to the tourists. This increases the revenue of locals as well. Tourists during their stay consume a variety of dishes. The following table lists the most frequently consumed dishes.

The most consumed dishes include Dham, Siddu, Madra, Pasta, Handi chicken, Sepu Baddi, Kadhi Rice, Rajma Rice, momos, and pizza. It is an ideal sign for the tourism industry as the tourists are consuming local dishes more than the dishes served by food chains. This adds directly to the economy of the state. It has implications for rural tourism development as well. As people are consuming local dishes, this means that they would love to eat dishes cooked traditionally. Therefore, the provision of cooking these dishes in their real ways should be made at rural destinations.

5.5.7 Knowledge of Dham

In our findings, Dham was identified as one of the dishes frequently consumed by the tourists. However, Dham is also served at religious places in Langars. So, many of the tourists who consume Dham are even not aware of what it is. In addition, the preparation of Dham differs from district to district. However, keeping everything in mind, Dham can be proclaimed a signature dish of the state and can be branded as the state dish to promote food tourism in the state. So for that reason, the opinion of tourists about their knowledge of Dham should be assessed. The tourists are inquired if they knew about Himachali Dham. The following responses were recorded.

In our survey, only 68% of the tourists knew about Himachali Dham. Even though the number is high, since our tourists comprise frequent visitors, 29% of unawareness about Dham is very high. This clearly shows that people are not much aware of the locally consumed dishes and give the real taste of the state's food culture. More than awareness, it is the consumption of Dham that matters more. The tourists were asked whether they have consumed Dham during their current or previous visit to the state. The following responses were recorded.

In our survey, it is revealed that only 57% of the tourists consumed Dham. This percentage is lower than the ones that are aware of it. It shows the lack of intent of tourists to consume it. This might be due to the low marketing of Dham or its unavailability. The percentage of tourists having not consumed Dham in their current or previous visits is very high. This shows that people are not well versed with the authentic food dishes of the state. The need for marketing of food products is highlighted with this analysis. By developing rural tourism, such authentic dishes and food products can get a platform for their promotion.

5.5.8 Food Recommendation and Its Consumption

It is the responsibility of other stakeholders to make tourists aware of the food specialty of their region. The tourism industry is a marketing-based industry. If the consumer is not aware of the available products locally, then there will be on sale for that. This will lead to economic loss for the industry. Our study revealed that other stakeholders suggested tourists about the local food products and dishes. About 76% of the tourists were suggested the dishes, and 90% of the tourists consumed the suggested dishes. This indicates that the marketing of these products will undoubtedly lead to increased sales for these products.

5.5.9 Services of Online Food Ordering, Cab Aggregator, Photographer, and Guide

One of the study's objectives is to tackle the socioeconomic challenges existing in the society with the help of rural tourism. Unemployment is one such challenge that exists not only in urban areas but in rural areas as well. Himachal Pradesh has a high unemployment rate which is the sixth highest in the country. The tourism and hospitality industry contributes directly as well as indirectly in generating employment. Be it the use of delivery representatives in the food delivery mechanism, employing people in cab services, or the need for trained and skilled personnel in the form of tourist guides and photographers, the industry's contribution in generating employment is very high.

In our survey, 86% of the respondents consumed their food by visiting the outlet. Only a meager 14% ordered on the online platforms such as Zomato and Swiggy. This high percentage is also because many tourists were staying in private hotels or outlets and therefore did not need to order from outside. In addition, since the number of people in a group was more, it was not suitable to order food from online platforms. Uncertainty about the food and outlet quality listed on these platforms also hinders food ordering.

This shows that tourists are not much comfortable in ordering the food online. Although there is a demand and high scope for employment generation in such a segment, without the supporting demand, the industry will not be able to sustain for long. The future of the employed people will remain uncertain.

The other segment where there is a lot of potential in tourist areas is cab aggregators' use in transportation. The use of services such as Ola and Uber makes it more convenient for the drivers and the consumer to avail such services. The rates are standardized, and there is also reduced fear of traveling to unknown destinations with unknown people. For drivers, it gets easier to attract tourists without the concept of bargaining. It also helps in developing a competitive advantage over other drivers. These platforms also offer the digital mode of payment, which also attracts users due to the convenience of traveling with less cash. However, connectivity issues pose a challenge for the users and drivers to sustain the system.

In our survey, it has been found that 76% of the tourists availed the services of local transport such as buses, autos, and taxis that were available at their disposal. This might be due to the network connectivity issues faced by the tourists. Moreover, since the number of cabs on the online platform is less, the tourists had used the readily available mode of transportation to avoid waiting in the queue. This has high implications for rural tourism development. Since the demand is not that much, not many cab drivers will be willing to list themselves on such platforms as it would not be economical for them. So, the policymakers will have to rely upon either the state-owned transportation mechanism such as buses or the private local vehicles. This will directly impact the cost of traveling of the tourists as private players will charge more than the government vehicles. Without regulations, the price gouging and surge pricing will be a common phenomenon.

5.5.10 Photographer/Tour Guides/Adventurers

Photographers and tour guides are critical areas in the hospitality industry that can absorb the surplus skilled and semi-skilled workforce. In uncertain times such as in the current scenario, generating employment for the government is an uphill task. Although it has an ability, the manufacturing sector cannot fulfill the need of the present times. Therefore, tourism industry development is crucial as it can accommodate a lot of local workforce in tour guides. This alternative is feasible for the government only if there is a demand for such services in the industry. As technology has arrived, it has begun to replace a lot of jobs in the industry. The use of high-end mobile phones and DSLR has reduced the need for photographers. Similarly, knowledge about the monuments and places is available all across the Internet. This has severely reduced the need for tour guides as well. The responses recorded in our survey are as follows.

The data also highlights the concerns discussed above. The number of tourists using and not using the services of photographers and tour guides is close. In addition, the majority of favorable responses recorded were about the use of adventurers. This tells us that a particular niche can absorb the workforce by generating jobs as a specialist with the help of technology alone is not possible for the tourists. So, the need for a specialist service provider should be identified in rural tourism, and people must be registered and trained to become one. This will help tackle the customer grievances and a better understanding of the customer's needs and wants. Moreover, in rural areas, people themselves wish to reduce the use of technology. Due to certain technological limitations, they will not be able to use existing technology. Consequently, the role of tour guides and photographers will be high. Thus, the jobs can be created in such a segment through rural tourism. However, the spotlight should be more on specialized services rather than the general one.

5.5.11 Experience with Local People at HP

The tourist satisfaction level is determined by multiple factors. One of these is the experience of tourists with the other stakeholders. The main component of other stakeholders in the local population is shopkeepers, restaurant owners, hoteliers, tour guides, photographers, and adventurers. If a tourist is satisfied with these people's behavior, it creates a trusting relationship among them. Tourist is more likely to consent with the suggestions given to it by people and might spend more freely on services. It also impacts the overall experience of the tourist.

Our study reveals that about 95% of the tourists are satisfied with the behavior of the other stakeholders. The primary reason for the dissatisfaction among a few tourists was the misbehavior of hotel staff, misguidance by some people about destinations, and over-charging by few people. This clearly shows that tourists have trust over other stakeholders. The issues recognized that hamper the relationship should be eliminated to feel substantially more certain during their stay and increase their visits.

5.5.12 Overall Experience of Their Stay

Tourist satisfaction is determined by their ranking for the overall experience. It also determines whether or not a tourist will visit again. An industry must work to increase the overall satisfaction level of the tourists. It includes the fulfillment of tour objectives, pricing, the existence of the desired infrastructure, problems encountered during the stay, and low deviations in expectations before the visit and actual experiences. The responses recorded in our study are as follows.

The analysis shows that the tourism industry can satisfy the tourists. About 93% of the tourists were satisfied with the industry experience. If we go further, 45% of the tourists are completely satisfied, 32% are moderately satisfied, and 16% are slightly satisfied. Only 2% of the tourists are extremely satisfied with their visit. As the satisfaction level of tourists is very high, so is their frequency of visit. Tourists will visit a destination till the time their satisfaction level is met. Thus, while developing rural tourism, policymakers must ensure that the tourists' satisfaction level should be positively met. Only then, the tourists will revisit the same or similar destinations. Failure to serve the tourists up to their expectation will lead to loss of tourists and economic gains.

5.6 Post-visit Analysis

This part of the chapter is dealing with the experience and assessment of tourists regarding the visited destinations.

5.6.1 Challenges Faced During Stay

Tourist satisfaction is a critical component of the industry. If the tourist is dissatisfied with any component during its stay, it leads to cognitive dissonance. A dissatisfied tourist avoids revisiting the destination and will create negative publicity for the destination by sharing reviews on social media platforms and blogs. This causes negativity about the destination in the minds of tourists and leads to economic loss. A tourist can face multiple challenges during its visit. Some of these challenges are due to natural causes, while some are caused by the other stakeholders' misbehavior. Natural causes are uncontrollable in nature and are unavoidable during any stay. However, other stakeholders' mismanagement can be avoided and must be resolved to enhance customer satisfaction.

Our survey revealed that natural phenomena causing distress to the tourists were water shortage; too unpredictable cool weather at times; connectivity issues in the network due to weather; accessibility block due to rainfall, landslides, or snow; and poor condition of roads and infrastructure. However, some respondents pointed out the lack of professionalism by hotel staff, parking issues, the problem of overtourism, traffic and road blockage and over-policing at tourist destinations, little use of technology at destinations, the problem of surge pricing at various places, misguidance by stakeholders, misinformation provided through social media platforms, the problem of eve teasing by locals, and lack of friendliness among locals. Such incidents reduce the satisfaction of the tourist. These issues are trivial for the development of rural tourism. Since the issues highlighted by the tourist will be present on a large scale in rural areas, therefore, the chances for higher dissonance in rural areas will be much higher. So, suppose issues of connectivity, accessibility, basic infrastructure, safety, and tourists' security, particularly for females, can be assured. In that case, only rural tourism could be a hit tourism product in the state.

5.6.2 Reasons for Liking or Disliking HP Environment

Tourists have several preferences that result in likeness or disliking to a particular event. It might be a case that a particular thing might cause pleasure to some and displeasure to others. Therefore, these events must not be rectified at an individual level. However, suppose an event causes displeasure to several tourists. In that case, that event might be regarded as a threat to the industry.

Our survey revealed that the primary cause of liking the state was its pleasant environment. It is a commonly known phenomenon that people prefer cool weather over the hot summer season. A lot of tourists visit hill stations to avoid the scorching heat of the summer. Pollution-free environment, the serenity of great Himalayan Mountains, beautiful destinations, variety of cultures, greenery and hills, cleanliness in the cities, and mental peace due to less crowded cities are the reasons people prefer to visit Himachal Pradesh. However, there were several reasons why the people did not enjoy their trip to the full extent. The use of plastic at tourist destinations and the resultant garbage were the reasons identified in our survey. People also did not enjoy the over-commercialization of activities at tourist destinations. Charging money for necessary activities also causes discomfort to the public. These results are significant as they tell us that people are comfortable with the state's environment and visit it to avoid the mental stress caused due to the overcrowded cities. Rural tourism is a product that incorporates this in its vision. By bringing people close to nature, it creates them a sense of relaxation and a peaceful mind. It generates positive vibes, which are a necessity in modern-day life. However, people discouraged the over-commercialization of activities at destinations. Therefore, it should be noted that certain activities of rural tourism should be free or bundled together with other activities and should not be charged.

5.6.3 Ecological and Social Environment Peculiarity

It is human nature to notice things that they are not comfortable with or find unique to their usual surroundings. These locals' habits can be a cause of appreciation by the tourists and can be adopted by them in their homeland or can cause a strange shock to the tourists as well. The indirect behavior exhibited by the other stakeholders has a lot of impact on the tourists as well. Even though the behavior is not aimed directly at tourists, it impacts tourists' minds about the other stakeholders. This creates a feeling of trust or distrust in tourists, which affects their relationship with the other stakeholders.

In our survey, it is seen that tourists were shocked to see the use of plastic by the locals in such a clean environment. Despite the fact that local people discover little issue with it, the expanding issue of trash and littering is currently noticeable. The problem of resource sharing and the inconvenience caused to the tourists is visible to the tourists. The locals' change in behavior toward tourists is also visible due to the increasing issues concerning tourism. Tourists were also not happy with the practice of throwing garbage into the river and water bodies that are a significant attraction for them. However, most of the respondents thought that the problem of littering and garbage at tourist destinations was caused by the tourists and not by the local people. Numerous respondents also noted that people in Himachal Pradesh have lived throughout their life following the principles of nature. They appreciate the local customs and values while living and enjoying their life sustainably.

Regarding the social concerns, the locals were found to be peaceful, soft-spoken, and well-mannered by the tourists. The tourists responded that although their income is based on tourism and people wish to increase their income, these locals should not overburden the nature to increase their income. Their concerns for nature are more than their economic gains. Since the study's objective is to provide insights for the development of rural tourism, the responses show that rural tourism development would not suffer from the stakeholders' side. There might be some objections from the locals if tourism violates their local regions. Still, the stakeholders' behavior will ensure that tourism is sustainable and causes more gains with minimum damage to the environment.

5.6.4 Suggestions for Sustainable Tourism Development

Suggestions from the tourists are very crucial for the policymakers. They act as guidelines for the policymakers to develop policy strategies for the tourism products. They convey the policymakers about the possible evolution of tourism products in the industry. It also helps remove the service deficits and encourage the tourists and other stakeholders to establish a relationship of trust among them. Suppose tourists find that their suggestions have been incorporated and the supply side takes care of their reasonable demands. In that case, they feel much valued and try to empathize with the other stakeholders as well. This is visible in the higher economic contribution made by them to the industry.

In our survey, the tourists gave several valuable suggestions that would help in rural tourism development and help in incorporating sustainability in the industry. The main suggestions included increasing advertising for tourism-related activities and providing infrastructure, enhanced use of social media to spread knowledge about the tourism products, improved garbage disposal mechanism and imparting knowledge about it to the locals as well tourists, improving the condition of roads, banning the use of single-use plastic, affordable transportation, providing at least necessary infrastructure at various destinations, making locals more aware about the hospitality conditions, improvement in logistics, and better safety to the tourists in the dark and hilly areas. While developing rural tourism, these suggestions should be incorporated. The presence of necessary infrastructure is a must for the tourists. In addition, increasing the problem of garbage and its disposal should be thought on. Locals should be well trained to have basic hospitality manners, and accessibility should be improved all over Himachal Pradesh by maintaining roads and affordable transport means.

5.7 Rural Tourism Knowledge

Before introducing rural tourism as a fully developed niche tourism product, the existing knowledge of tourists must be assessed. This will additionally help to identify the tourist's hidden aspiration from the product and ascertain changes to the thought-upon and designed tourism products so as to meet the tourist expectations. Rural tourism is described as the tourist visits to a rural area for active participation in a rural lifestyle. It is described as a travel activity in the premise close to the rural region. It includes consuming nature-based activities such as hunting and fishing, agriculture produce, walking in the jungle, desert safari, living local lifestyle, adopting local customs and traditions, consuming local cuisines, using local handicrafts, enjoying local festivals, and living and residing in local ambiance rather than in a hotel or a resort.

However, the results in our survey are not that encouraging. Only limited respondents could answer the questions concerning rural tourism. The majority of the tourists had a vague idea about the concept of rural tourism. Their knowledge about the topology is limited only to visiting the rural areas and villages. Some of the respondents also defined it as tourism undertaken without much use of technology. None of these responses define the rural tourism in their true essence. This outcome is problematic for the policymakers. The absence of knowledge about this typology indicates that there has been pretty much low advertising about this particular product. Prior to launching the product, first, tourists have to be made aware of the concept of rural tourism and the activities that could be performed under it. Additionally, the advantages from rural tourism and the need for developing rural tourism should be known to the tourists. However, the responses also throw some light on the desires of tourists from rural tourism. Through assessment, they clarified that the presence of only indigenous goods, food, and festivals is a must for rural tourism destinations. Further, the role of technology should be minimum at such places. The destinations developed should incorporate visits to ancient monuments, historical sites, sites with cultural and religious importance, farmlands, and organically grown food and crops. The respondents likewise proposed that in this tourism typology, the participation of tourists in rural people's activities is a must. Therefore, while choosing and developing rural tourism destinations, tourists' vision should also be kept in mind.

5.8 Wish to Visit Rural HP

Not only has the knowledge about a product counted but the intention to consume it as well. Despite the fact that the consumer is unaware of the product and intends to consume, it is sufficient to conclude that the product shall have a market demand if launched strategically. As the tourists are willing to increase their stay, they can divert them to rural destinations. However, the opinion of the tourists should be assessed in this regard. There are possibilities that tourists would not want to visit the destination and continue with the existing plans or might deviate a little bit from their plans. This is on the grounds that tourists do not generally prefer to venture into the unknown if they travel in a group or with their family. In our survey, we asked the tourists whether they would like to visit the rural destination of Himachal Pradesh. The responses recorded are tabulated in the table below.

In our survey, the tourists are asked if they will switch their current tour or plan and travel to the state's rural destinations (Table 5.14).

The results show that around 90% of the tourists are willing to experience the state's thrill of rural destinations. Around 9% of the tourists were not confident enough because of lack of awareness about the product, uncertainty about the availability of desired infrastructure, and security concerns. Only a handful of respondents declined to visit the rural destinations for tourism purposes. The results here

Table 5.14 Wish to revisit Himachal Pradesh

Response	Frequency
Yes	362 (90%)
No	3 (1%)
Maybe	35 (9%)
Total	**400**

Source: Primary survey

are encouraging for the industry that people have no hesitation in visiting the state's rural destinations. This means that rural destinations, if appropriately marketed, can attract a lot of tourists. However, the results should not be concluded that tourism will boom as well. Mere affirmation to visit the rural areas' destination does not mean that people will consume the rural tourism products. But, the intention of the tourist poses a favorable situation for the policymakers.

5.9 Essential Infrastructure

The essential condition for attracting tourists is the presence of the necessary infrastructure. Basic infrastructure includes the presence of rooms and hotels with proper bedding and sanitation, hygienic facilities, network availability for communication, accessibility of destinations through roads, affordable means of transport, availability of a variety of foods, and presence of marketing for shopping.

In our study, the respondents opined that cleanliness is one of the most critical infrastructure characteristics present at the tourist destinations. Further, there is no requirement for high-tech communication networking. Still, the presence of a mobile network is a must for communication. There should be a map available at the district or block headquarters available for the tourists. Some of the respondents suggested the need for photographers, tourist guides, and adventurers at the adventure site. This indicates that if rural tourism is to be developed, then these things should be present at such destinations. This survey also highlights the potential of rural tourism development in creating employment opportunities in rural Himachal Pradesh. What is needed here is ensuring skilled or semi-skilled people who can work as a photographer or tour guides. Local people who are educated but are working in agriculture can be encouraged to take up the jobs and move out of poverty's vicious circle. Additionally, the predicament of infrastructure—tourist arrival, which is common while developing destinations—can be resolved as the development of only basic infrastructure will not cause much damage to the environment and is a cost-effective measure. The infrastructure developed was utilized by local people to help in its maintenance as well.

5.9.1 Hygiene

A tourist destination must be hygienic. It is the first and topmost priority of any tourist while selecting destinations, hotels, restaurants, and other similar services and products. It is one of the components of the necessary infrastructure needed by the tourist. It is also essential for the survival of the tourism industry. The need for hygiene is also independent of age group and gender. It is required by every age group and every gender. The importance of hygiene increases if there are senior citizens in the group or there are small children. The analysis of data collected from the tourist also highlights the importance of hygiene at tourist destinations (Table 5.15).

Table 5.15 Tourist preference for hygiene

Responses	Frequency
1. (Extremely unimportant)	11 (3%)
2. (Moderately unimportant)	8 (2%)
3. (Slightly unimportant)	4 (1%)
4. (Neutral)	21 (5%)
5. (Slightly important)	36 (9%)
6. (Moderately important)	98 (25%)
7. (Extremely important)	222 (56%)
Total	**400**

Source: Primary survey

In our survey, around 90% of the tourist responded that hygiene is an essential aspect of infrastructure at a destination. The hotel rooms must be clean and have proper-functioning washroom, availability of garbage bins, and dustbins at destinations and hotels; hygiene in restaurants is an absolute necessity for the tourists. So, when the destinations are developed in rural places, they have to be kept in mind. The rooms or cottages developed need to have proper illumination, ventilation, and proper room maintenance. In addition, since it is the need of every age group, therefore, a standard should be set for hygienic activities that should be followed at every such destination.

5.9.2 Technology

Suppose some factors are essential and are needed by the tourist to be available at destinations compulsorily. In that case, there is an infrastructure that is termed as luxury. This includes all those things that a tourist desires to be present at the destination. Its satisfaction level rises. The shortfall of these factors does not cause dissatisfaction among the tourists. In our study, technology is revealed to be one of the foremost luxury infrastructures. It includes Wi-Fi in hotels and hotspots at tourist destinations, availability of television and cable network in hotels, availability of digital payment mechanism at several possible places, mechanism of hot and cold water in washrooms in hotels, and the facility of luxury dining and theme-based hotel rooms. Albeit these desires are considered a luxury today, but as the tourism industry is evolving, there are chances that these might be a part of the essential infrastructure at tourist destinations. These services can act as a competitive advantage while developing rural tourism destination sites. By ensuring the presence of Internet connectivity or theme-based dining, tourists can be attracted to such places.

5.10 Conclusion

This section introduced the analysis of tourists visiting Himachal Pradesh and their implications for the development of the rural tourism industry while increasing the economic opportunities (Ashley et al. 2007). The chapter concludes that the seasonality of visits has a significant relationship with the average duration of stay. So, while developing rural tourism, the season for promoting and encouraging it should be where the length of stay of tourists is less. In the seasons where the average duration of stay is higher, due to the nonavailability of hotels, tourists will be automatically diverted to the rural destinations without the need for marketing. In addition, the income of tourists has a significant relationship with the frequency of visits. So, tourists with less income groups should be attracted to rural tourism. High-income groups will like to visit again but with increased desires and expectations. So, this will put a mire burden on destination development. The age group of the tourists also has a significant relationship with the frequency of visits. So, the age group that visits less frequently should be targeted for rural tourism to avoid unnecessary product modifications. This will have no impact on the earnings as each age group accepts deviations in their budget. So the economic gain from different age groups is relatively similar. The analysis shows that there is scope for rural tourism development. However, substantial marketing efforts will be needed to make people aware of it. Also, informative content about rural tourism should be displayed across social media and over the Internet so that it is easier to access. It is to be ensured that proper social media management is undertaken to help tourists make rational decisions for their trip. Negative reviews must be countered and resolved so that the image of the tourism product is not diminished. Further, as rural tourism will be a less expensive affair for the tourist, the tourist will be happy to stay for more without compromising much on its budget. To develop rural tourism sustainably, the destinations must be identified in the districts with an abundance of natural resources, necessary infrastructure, and less tourist footfall. The chapter also concludes that to make rural tourism successful, it has to make a part of a package to search about it and avail it before starting their tour. Rural tourism also has enormous potential in generating employment opportunities in rural areas. This will help in tackling several socioeconomic challenges about the state. The analysis shows that tourists have trust over stakeholders, but misbehavior also causes extreme dissatisfaction. Such actions hamper the prospects of the development of the industry. Overall, the tourists are satisfied with the industry and therefore likely to visit the state frequently. In a nutshell, the analysis of tourist behavior poses favorable prospects for developing the rural tourism industry in Himachal Pradesh.

References

Ashley, C., P. Brine, A. Lehr, and H. Wilde. 2007. *The role of the tourism sector in expanding economic opportunity*. Corporate Social Responsibility Initiative Report No. 23. Kennedy School of Government, Harvard University, Cambridge.

Bird, S.R., and S.G. Sapp. 2004. Understanding the gender gap in small business success: Urban and rural comparisons. *Gender and Society* 18 (1): 5–28.

Bramwell, B. 1994. Rural tourism and sustainable rural tourism. *Journal of Sustainable Tourism* 2 (1–2): 1–21.

Cole, V., and A.J. Sinclair. 2002. Measuring the ecological footprint of a Himalayan tourist center. *Mountain Research and Development* 22 (2): 132–141.

Dreze, J., and A. Sen. 2002. *India: Development and participation*. Oxford: Oxford University Press.

Kumar, R. 1993. *The history of doing: An account of women's rights and feminism in India*. New Delhi: Zubaan.

Ngoc, N. M., Tien, N. H., Hieu, V. M., and Trang, T. T. T. 2023. *Sustainable Integration in Vietnam's Tourism Industry*. World review of entrepreneurship management and sustainable development.

Shrivastava, A., and M. Heredge. 2004, January 8. Rural tourism in the Seraj Valley of Himachal Pradesh.

6 Perception of Other Stakeholders on Development of Rural Tourism in Himachal Pradesh

6.1 Introduction

It is a common saying that supply creates its own demand. In the event that the demand side's perception is critical to know about the new product's success, then the opinion of the supply side matters too. The availability of the product and how it is provided determines how it will be received by the consumers. The tourism industry's supply side comprises hoteliers, restaurant owners, travel agents/tour operators, homestay owners, photographers, tourist guides, adventurers, and other similar persons (from here on, referred to as other stakeholders). The table below shows the categorization of other stakeholders based on their role in the industry and their operation district (Table 6.1).

Till now, planning in the tourism industry is done without any participation of stakeholders. These inflexible plans, accordingly, neglect to attract the interest of other stakeholders. Without their required cooperation, the plans lose its effectiveness and fail to achieve the desired objectives. The policymakers are often criticized by the researchers and academicians because of their lack of empathy while policymaking. Their rigidness in including the stakeholder's perspective has caused much more damage than benefit. Therefore, to avoid a similar fate for the rural tourism industry, this study makes policymaking more inclusive. By inviting suggestions and incorporating them into the strategy, the stakeholders are given their due importance. This way, it gives them a sense of belongingness to the product developed. All things considered, it likewise enhances the overall output achieved. By tuning in to the issues of these stakeholders, tourism products can be made more stakeholder friendly.

S. Kumar, *Sustainable Rural Tourism in Himalayan Foothills*,
https://doi.org/10.1007/978-3-031-40098-8_6

Table 6.1 Descriptive statistics of the respondents (other stakeholders)

S. No	Demographic factors	Variables and percentage
1.	Role in the industry	Hotelier/guesthouse owners (28%)
		Homestay owners (12%)
		Restaurant owners (7%)
		Travel agents (14%)
		Photographers+ adventurers (11%)
		Tour guides (9%)
		People representative (8%)
		Cab/taxi operators (13%)
2.	District of operation	Bilaspur (2%)
		Chamba (5%)
		Hamirpur (1%)
		Kangra (13%)
		Kinnaur (1%)
		Kullu (34%)
		Lahaul & Spiti (3%)
		Mandi (5%)
		Shimla (24%)
		Sirmaur (3%)
		Solan (7%)
		Una (1%)
		n = 400

Source: Primary survey

6.2 Results and Discussion

To analyze the data and discuss the results, the responses gathered are categorized based on their job nature. Similar jobs are grouped. The conclusion has been derived from the data collected. The analysis done will help us assess the perception of other stakeholders involved in the supply side of the tourism industry about the possibility of developing and sustaining the rural tourism in Himachal Pradesh. The following section presents a detailed analysis of the perception of different stakeholders.

The analysis is done through qualitative techniques. The nature of the analysis is exploratory. It gives a detailed view of perception with an explanation for the same. The analysis is done using qualitative data analysis software NVivo. The results are divided into several segments. Each segment details the various aspects related to a particular stakeholder. These aspects include their earning, previous occupation, use of technology, modes of marketing, challenges faced by them, tourists catered, their perception on rural tourism development, hindrances in rural tourism development, their willingness to operate in rural areas, the difference in products offered in urban and rural areas, and others. Thus, it details a complete exploration of the various aspects of stakeholders in the tourism industry (Table 6.2).

Table 6.2 Responses of hoteliers/guesthouse and homestay owners on different questions

Particulars	Response
Season of the tourism	Peak season (April to June)
	Lean season (July to September)
Occupancy during seasons	Peak season (100%)
	Lean season (25–35%)
Average occupancy rate	65–75%
Destination recommendation	Yes (n = 135)
	No (n = 25)
Food recommendation	Yes (n = 120)
	No (n = 40)
Scope for rural tourism	Yes (n = 145)
	No (n = 8)
	May be (n = 7)
Willingness to work in rural area	Yes (n = 132)
	No (n = 15)
	May be (n = 13)

Source: Primary survey

6.2.1 To Visualize Cumulative Response of Hoteliers/Guesthouse and Homestay Owners

Hotels, guesthouses, and homestays are the prominent players in the tourism industry. They cater to approximately all the individuals visiting the state directly or indirectly. Every tourist needs a place to stay. This segment of stakeholders creates a first impression on the tourists and determines the overall satisfaction for them. Besides providing a place to stay, it provides tourists all the relevant information that assists them during their travel. For the development of rural tourism, necessary infrastructure must exist in that place. Neat and clean rooms and washrooms are necessary infrastructure that is essential for the tourists. It is also a must that stay rooms in rural areas are cost-effective to not make the travel costly. Because of less economic profits, there are chances that the industry might not be willing to invest in such a risky venture, so it is crucial to have their perception before proceeding with the development.

Peak Season and Lean Season

The results obtained from hoteliers and homestay owners indicate that the summer season, i.e., April to June, is the peak season of tourists' arrival. The hotels during the peak season are completely occupied and operate at their 100% capacity. There is a shortage of rooms available in good hotels, and people are willing to pay more to get the rooms. However, during the lean season, the occupancy rate falls to around 25–35%. During such times, the hotels offer discounts to grab tourists.

6.2.1.1 To Discern Occupational Preference During Lean Season

The primary occupation that emerged as an alternative for such people is farming activity and engaging in daily activities. Since the demand is low during the lean season, the staff is encouraged to work on non-tourism activities such as farming so that their livelihood should not be severely impacted. As the season arrives, the staff engaged is called back to perform their respective task.

6.2.1.2 To Observe Average Tourist Duration of Stay

It is observed from the data that the average duration of stay of tourists in around 4–5 days. The average duration for the tourists staying in hotels is lower (2–3 days) than the tourists staying in homestays (4–5 days). This is because of the higher rent of the hotels and restrictions on the number of people that can accommodate in a single room. Since homestays pose no such restriction, people traveling in groups prefer to stay in homestays.

6.2.1.3 To Overcome Language Barriers with International Tourists

Hoteliers avail the services of local tourist guides to remove the hindrance of foreign languages with international tourists.

6.2.1.4 To Find Out Basic Services in Room and Extra Demands of Tourists

It is observed that the primary demand of tourists includes a neat and clean room and hygienic food. In addition, a clean washroom is one of the most essential demands for the tourists. Tourists also demand toiletries, hot water, and electric kettles in their rooms to avoid any inconvenience. Additional demands of tourists include the availability of liquor in their hotel room, Wi-Fi availability, and a working television with national TV channels and air conditioners in summer seasons. Individual tourists demanded rooms that give a scenic view from their window.

6.2.1.5 Adopted Marketing Strategies for Tourists

Hotels and homestay owners have found a variety of ways to market themselves to tourists. It includes marketing on social media platforms such as Facebook and YouTube of famous travel influencers, building their websites; tying-up with other platforms such as MakeMyTrip, Trivago, Goibibo, etc.; services of travel agents; and collaborating with tour guides, taxi/cab operators for commission.

6.2.1.6 Recommendations on Destination and Food Provided to Tourists

The tourists are suggested food and destinations as per our survey. However, whether or not tourists consumed the food recommended and visited the destinations is unknown to the hoteliers and homestay owners.

6.2.1.7 Follow-Up on the Tourist's Feedback

In our survey, the hotels revealed that they are very strict on the tourists' recommendations, suggestions, and complaints. This is because any lapse in service might trigger social media disaster, which could harm the hotel's image or homestay. All of the respondents agreed that they ask for feedback from the tourists during their checkout and work. A hotel revealed that once a tourist suggested a separate theme-based dining area to give it a more romantic and private feel for the tourists visiting the destination with their loved ones. The hotel is currently working on the suggestion as young couples do frequently visit their destination. Similarly, due to the high frequency of religious tourists, certain hotels have started preparing sattvic food separately from non-vegetarian food because of suggestions made by the number of tourists. Hotel managers revealed that by working on their suggestions, the hotels tend to make tourists feel more valuable and create a sense of belongingness and reliability to them.

6.2.1.8 Usage of Technology

The use of technology in accepting payments was limited to big private hotels. Government guesthouses operate solely on a cash basis. Homestays in various areas are also reluctant to accept digital payments because of various technical glitches. From this, it can be concluded that this segment does not use technology to a large extent. However, customers were allowed to book their rooms on online platforms and avail of discounts and e-gift vouchers for their next visit. With this implementation, certain private hotels can attract more tourists than small hotels.

6.2.1.9 To Find Out Associated Problems with Tourism Industry

The major problem revealed by the concerned people is the lack of government support for the hotels. There is a considerable funding problem for the small hotels that operate in the state. Without adequate funding, they are unable to compete with the big hotels in acquiring the necessary resources. In addition, the quality of their service is substandard when compared to big hotel chains. Homestays do not suffer from funding issues because of the government's subsidy under their schemes.

6.2.1.10 View on Scope of Rural Tourism

On discussing the potential for developing the rural tourism industry in Himachal Pradesh, a staggering 91% of the hoteliers and homestay owners confirmed in affirmative. The reason for such favorable figures is the presence of various unexplored tourist destinations in rural regions and the vast rural landscape in Himachal. The respondents suspecting the development believe that their districts do not have anything exclusive in rural areas that could attract tourists. The existing products, according to them, include all the things that rural tourism will incorporate.

6.2.1.11 Opinion on Willingness to Work in Sector of Rural Tourism

Assume the hoteliers accept that there is potential for the development of the rural tourism industry. In that case, they must be willing to operate their services in this area. This is because the objective of the industry is to earn profit from customer satisfaction. If an industry is being developed, the presence of other stakeholders is crucial for it. So the hoteliers were enquired about whether they will be willing to operate in rural areas if the industry is developed. Of the responses collected, 83%

agreed to the proposal. However, 10% did not agree that the proposal cited low economic gains and difficulty maintaining their standardized service quality as the primary reason. The hoteliers believe that the pricing of their rooms will be much lesser in the rural areas. So, the revenue earned from it won't be adequate to continue operating the services at their standardized service quality.

6.2.1.12 Product Preference Offered in Rural Areas

Urban hotels will be quite different from the hotel rooms present in rural destinations. This is because hotels' main objective at such a destination will be to provide the tourists with the necessary infrastructure. Bamboo hut-type rooms will be the most preferred type of rooms constructed in such areas that give an authentic feel to tourists. There will be fewer technical components, such as the absence of Wi-Fi and television sets in the room. However, for water serving, use of earthen pots might be considered. Food served in hotels will be more like local cuisines cooked in authentic style. Apart from this, the pricing per room will be lower than the room charged in urban hotels.

6.2.1.13 Challenges Faced in Development of Rural Tourism

Of the major challenges cited by the hoteliers, the prominent ones include supporting infrastructure such as roads, communication networks, and underdevelopment of the tourism industry in rural areas. The hoteliers and homestay owners believe that rural tourism will not be a lucrative typology in the absence of integrated and persistent marketing efforts by the industry. The presence of guides, tour operators, photographers, and non-tourism stakeholders such as hospitals, eating joints, and shopping markets will be a major challenge for the industry. The local population also exhibits a crucial impact on the choice of tourist destination as per hoteliers. So, the uneducated population in rural areas will hamper tourist satisfaction. Living standards of rural areas might also be a challenge for the incoming tourists.

The above discussion highlights critical takeaways for the policymakers for the rural tourism development. First, hotels' pricing in such a place will be lower than the current rates at urban destinations. So the economic gains from the trip will be lower than the urban trips made by the tourists. Second, a vast majority of the hoteliers and homestay owners are willing to work in rural areas; however, they might need financial assistance. They would not be able to maintain their quality of service. Third, the hotels and homestay products will be much different from what they are currently offering at urban destinations. With much focus on necessary infrastructure in hotels, what is required is the supporting industry. Thus, the presence of electricity, right roads, and mobile communication networks should also be arranged for the tourists (Table 6.3).

6.2.2 Response of Restaurant Owners

Apart from a place to stay, hygienic food is what a tourist requires while visiting a destination. Food is a significant component of traveling. It can fabricate a memory

Table 6.3 Responses of restaurant owners on different questions

Particulars	Responses
Food license	Yes ($n = 25$)
	No ($n = 2$)
Do you provide home delivery	Yes ($n = 7$)
	No ($n = 20$)
Open restaurant in rural areas	Yes ($n = 23$)
	No ($n = 2$)
	May be ($n = 2$)
Accept digital payments	Yes ($n = 21$)
	No ($n = 6$)
Preferred mode of payment	Cash ($n = 18$)
	Digital payment ($n = 3$)
Provide home delivery through self or digital platforms	Yes ($n = 22$)
	No ($n = 5$)
Orders received on digital platforms	0–4 ($n = 2$)
	5–9 ($n = 8$)
	10–14 ($n = 10$)
	15–19 ($n = 1$)
	More than 20 ($n = 0$)

Source: Primary survey

of a taste that can compel a tourist to visit again or remember and publicize it for a lifetime, or it can spoil a trip and be remembered as a bad memory for ages to come. Himachal Pradesh has many registered and unregistered restaurants running across the state. They serve traditional dishes like Dham, Siddu, and Madra; North Indian dishes like Paneer, Naan, and chicken; south Indian cuisines such as idli, dosa, and sambhar; as well as western cuisine such as pasta, pizza, and sandwiches. The presence of big food chains also gives tough competition to such a restaurant in their operations. For the study, the perception of restaurant owners concerning rural tourism development is also analyzed. This is because a significant portion of tourists like to eat at places that are nearby. People eat at their convenience during traveling, so restaurants offering a variety of dishes hygienically are essential for developing rural tourism.

6.2.2.1 Availability of Food License

Allotment of food license is mandatory at several places to ensure that a reasonable standard of quality and hygiene is maintained at restaurant and food eateries. People do prefer to eat at places which are registered. While ordering food online as well, people do check if the outlet is a registered one or not. Of the restaurants in our sample, 92% of them has a license to operate, and only 8% is unlicensed.

6.2.2.2 Preferences to Offered Cuisine

Restaurants offer a wide range of food products. Be it a combination of north Indian dishes, south Indian dishes, continental dishes, traditional Himachali food, and

Table 6.4 Restaurant types and dishes served

Restaurant type	Cuisines served	Major dishes
Exclusive	Italian	Pizza, lasagne, risotto, tiramisu
	South India	Dosa, Uttapam, Upma, Idli
	Street food	Momos, Babru, Siddu, Khatta
	Chinese	Momos, Thukpas, Tingmo, soup
	Beverages	Sura, Angoori, Chulli, Behmi, arak
	North Indian	Kadai paneer, chicken, mixed veg
Regular: Multi-cuisine	North Indian, Himachali food, South India, Italian, street food, Chinese	Kadai paneer, chicken, mixed veg, Dosa, Uttapam, Sura, Angoori, Siddu, Khatta
Pure vegetarian	Sattvic food	Dal makhani, mixed veg, and butter roti
Specialty restaurant	Vegan food	Buddha bowls, soups, and curries
Religion based	Jain restaurant	Paneer, ice creams, lotus stem
	Buddhist restaurant	Buddha bowls, soups, and curries

Source: Primary survey

Italian cuisines. There are exclusive outlets of big food chains such as Pizza Hut that serve only a particular food type. The majority of the restaurants, however, served a combination of different food cuisines (Table 6.4).

The USP dish differed widely in each of the restaurants. While analyzing the USP dish, exclusive, specialty, and religion-based restaurants are excluded. They will give results that will be biased. From the data obtained from the regular multi-cuisine restaurant, it is observed that butter chicken in the north Indian category and Dham and Siddu from the traditional Himachali food category emerged as the winners. Since chicken is not the state's specialty and is commonly found across the country in every state, traditional food items like Dham and Siddu can be attributed to the USP dish of various restaurants.

6.2.2.3 Average Table Occupancy Observed During Peak Season

The average occupancy in restaurants during the peak season varied based on the cuisines they serve. Exclusive multi-chain outlets like Pizza Hut have around 70–80% of the occupancy during the peak season. Multi-cuisine regular restaurants have around 85–95% of occupancy during the peak season. In addition, the place where the restaurant is located plays a pivotal role in deciding the occupancy. Restaurants that are close to the famous destinations have a higher occupancy rate than the restaurants that are not easily accessible. However, if situated within the same locality, then multi-cuisine restaurants have a higher occupancy rate than the exclusive restaurant. It is also essential to understand that the occupancy rate was asked in terms of tourists only. The local population is not accountable in the occupancy rate determined.

6.2.2.4 Occupancy Observed During the Lean Season

During the lean season, the occupancy rate at the restaurants steeply declined. Multi-cuisine regular restaurants have around 30–40% of the occupancy, whereas exclusive chain-based stores show an occupancy rate of 25–30%. During the lean season, more locals visit these restaurants instead of the tourists themselves.

6.2.2.5 Average Customer's Bill Amount Received at Restaurants

The average expenditure made by the tourists depends largely on the dishes they consume at the restaurants. Exclusive chain-based stores offer various discounts and offer products at prices that are nationalized and fixed by the franchisee. The average amount of bill differed on the number of cuisines served by the restaurant. The table below lists the average bill amount that a consumer pays while visiting (Table 6.5).

6.2.2.6 Availability and Ease of Free Home Delivery

Home delivery of food is also another way restaurants try to edge out their competitors. By delivering food to the nearby homestays, hotels, or destinations, restaurants can attract tourists toward them. The service is complementary and free of charge. Sometimes, a delivery fee is chargeable based on the distance covered or the amount of order.

As per the data collected, most of the restaurants surveyed do not provide the service of home delivery and are dine-in or takeaway type of restaurants. The reason cited for the same is the preliminary order received for the delivery to meet hiring a person for the same. However, a new type of delivery system is also emerging in the state. Food aggregators such as Swiggy and Zomato have popped up and handle the delivery of orders placed on their application.

Table 6.5 Cuisine served and average bill of customers

Restaurant type	Cuisines served	Average bill (for 2 persons)
Exclusive	Italian	Rs. 1000–1200
	South Indian	Rs. 400–500
	Street food	Rs. 100–150
	Chinese	Rs. 300–350
	Beverages	Rs. 250–350
	North Indian	Rs. 400–500
Regular: Multi-cuisine	North Indian, Himachali food, south Indian, Italian, street food, Chinese	Rs. 300–350
Pure vegetarian	Satvik food	Rs. 500–600
Specialty restaurant	Vegan food	Rs. 800–1000
Religion based	Jain restaurant	Rs. 400–500
	Buddhist restaurant	Rs. 600–700

Source: Primary survey

On enquiring if the restaurants have tied up with such food aggregators, 84% replied in the affirmative. This means that there is scope for the development of an online food ordering system in the state.

The number of orders received on such platforms ranged from 10 to 14 orders per day. This is a good number of orders, but these also account for the local people's orders. The number of orders is also high because of the increased promotional discounts offered by such platforms in the initial operation stages. The restaurant owners also revealed that people are sitting inside a mall place order on the platforms to avail discounts rather than placing directly from the shop located in vicinity. These practices usually make it challenging to imply concrete results.

The most ordered dish from the restaurants is pizza and chicken. Street food items such as momos, rolls, and tea are preferred to be eaten at the place of offering as delivering these items may result in loss of taste due to packaging and spillage. In addition, the limitation of packaging poses restrictions on the items that can be delivered. Interestingly, traditional food items like Dham, Siddu, and Madra did not have any takers on such platforms. This is because tourists might not be aware of them directly and needs to be marketed by someone to convince them to consume. These dishes portray not much of interest for locals, and therefore they do not order such dishes on online platforms.

6.2.2.7 Employed Marketing Strategies to Allure the Customers

No business can survive without marketing. For marketing, multi-chain franchisee-based outlets survive on listing on parent website, placing billboards on bus stations and malls, reaching out to customers using their social media accounts. Restaurants are also tied up with tour guides and taxi/cab operators on a commission basis for bringing tourists to their restaurant. Most of the restaurants, however, have no such distinct marketing plan for them.

6.2.2.8 Acceptance to Digital Payments

Digitalization is one way in which the industry is experiencing some tremendous challenges. The efficiency of industry in serving tourists is increasing at a rate higher than before. Accepting digital payment is one of the most convenient ways in which the industry has adopted digitalization. Our survey revealed that 78% of the restaurants accept money in digital form through UPI platforms, credit/debit cards, or the use of e-wallet.

As there is provision of making payments digitally, tourists' preferred mode at such places is also digital payment. This is due to the ease of traveling without thinking about money. The tension to order the dish that fits in the amount of money available in cash is eased out. The digital payment sites' further discounts also make it much more attractive to use than the cash system.

6.2.2.9 Measures to Overcome Language Barriers with International Tourists

Language barriers with international tourists are common problem in the hospitality industry. The restaurants have adopted several measures to tackle the same. They have employed a person who can speak and understand English as most international tourists have some basic knowledge of English. Further screen-based menu is used in high-end restaurants through which tourists can translate the menu using their digital devices. This way, both the tourists and the restaurants can bridge the linguistic barrier between them.

6.2.2.10 Feedbacks Observed and Suggestions Incorporated Regarding Food Offered to Tourists

All of the restaurants surveyed have a means or other to obtain feedback from the tourists served. Some use a feedback form for the same, while others ask tourists to provide feedback on their website or social media pages. Suggestions are equally welcomed by the restaurants for improvement. One of the restaurants revealed that once, an international tourist asked for a dish prepared in an authentic Himachali style. As the dish is spicy, the chef decided to undertone the spiciness level, keeping in mind the foreigners' taste. However, the tourist was not satisfied with the dish. It was below his expectation formed through reviews on social media. Thus, the restaurant and the chef decided not to interfere with the recipe as tourists have expectations based on some research over the Internet.

6.2.2.11 Problems Faced by the Restaurant Owners

The restaurant industry suffers from a lack of tourist footfall, increasing competition, increasing the cost of operation, and mushrooming up of multi-chain outlets in the state. To beat the competition, the price war is inevitable, which results in economic losses for the industry. In addition, poor marketing makes it much more troublesome for the restaurant owners. They have to fight with the big established brands with deep pockets that have standardized food product offerings. Moreover, hotels these days have food charges included in their stay packages. This restricts tourists from eating outside. Thus, we can imply that the restaurant industry needs more tourists to survive.

6.2.2.12 Views on Willingness to Open Restaurant in Rural Himachal Pradesh

Since the lack of tourists is one of the major concerns for the industry and increasing environmental concerns have restricted the number of tourists, there is a need to develop tourism products at destinations that still need to be explored. One way is to move toward the rural regions, which have resource abundance and less population density and can be used to attract tourists. The restaurant owners were asked if they will open a branch of their restaurant in rural areas if rural tourism is developed. The ensuing table delineates the documented responses:

The response recorded reveals that restaurant owners are willing to move to the rural areas well if the demand arises due to rural tourism development. However, they revealed that the products they are currently offering will not be the same as the products offered at rural tourism destinations.

6.2.2.13 Differential View on Dishes Served

On enquiring about the difference in products offered by them, the restaurant owners revealed that the products such as western cuisine and continental food will not be a suitable cuisine at rural destinations. In addition, the way food is being cooked will be drastically different. The spices used and preparing and cooking will be more traditional to give authentic food. The dishes served will be authentic local food eaten by people in rural Himachal in their daily lives. As far as cost is concerned, prices will be lower than the food prices in urban places because of low operational costs and the easy availability and procurement of raw materials.

The above discussion highlights critical takeaways for the policymakers for the rural tourism development. First, food dishes in a restaurant at rural destinations will be lower than those at urban destinations. So the economic gains from the tourist will be lower than the urban trips made by the tourists. Second, there is a lack of marketing efforts made by the restaurant industry as people get attracted easily to the multi-chain outlets mushrooming up in the society. The easing of competition through opening new avenues will be highly appreciated by the industry. Third, the restaurant's products will be much different from what they are currently offering at urban destinations. With much focus on traditional authentic cuisine, the mere presence of limited tourists will do so for the industry in the initial stage.

6.2.2.14 Perception Toward Photographers and Adventurers

Capturing memories is one of the critical components of a trip. Tourists want to capture their emotions, feelings, and excitement for a long time. As social media is engulfing our lives, the trend of uploading high-quality images to get likes and reactions is increasing day by day. Although the use of smartphones and high-end cameras by tourists themselves has started, people wish to engage a photographer to capture images. This way, they can enjoy freely without worrying about the gadgets and distracting themselves from mesmerizing emotions. Adventurers, on the other hand, work to ensure safety and enhance tourist satisfaction in adventure sports. As the state is famous for adventure tourism, the role of adventurers is significant. Since tourists always wants to capture memories everywhere, rural tourism will not be an exception. Therefore, the assessment of the perception of photographers and adventurers concerning rural tourism development is essential.

6.2.2.15 Marketing Prospects for Customers

Marketing is an essential aspect for photographers. They have created various social media accounts, where they try to attract prospective tourists by showcasing their photos and recorded videos. They have also tied up with hotels, homestay owners, and travel agents that recommend them to the tourists coming in their vicinity. Adventurers register themselves with the government and provide services in their allocated areas. As they require professionalism in their service and have to follow certain safety precautions, along these lines, they pitch themselves through their professionalism and customer satisfaction reviews. They also tied up with hotels and homestay owners, who recommend their services to the tourists.

Table 6.6 Age-wise classification of photographers and adventurers

Age group	Photographers	Adventurers
Below 18 years	00	06 (28%)
18–25 years	02 (9%)	12 (58%)
25–35 years	08 (36%)	03 (14%)
35–45 years	08 (36%)	00
More than 45 years	04 (19%)	00
Total	**22**	**21**

Source: Primary survey

Table 6.7 Average earnings (per tourist) of photographers and adventurers

Earnings (in Rs)	Photographers	Adventurers
1–500	06 (27%)	00
500–1000	10 (45%)	05 (21%)
1000–1500	03 (14%)	08 (39%)
1500–2000	02 (9%)	07 (34%)
More than 2000	01 (5%)	01 (5%)
Total	**22**	**21**

Source: Primary survey

6.2.2.16 Services Catered Across Various Age Groups of Customers

While certain aspects in touring like boarding lodging and food are common for everyone, i.e., these are consumed by everybody, photographers and adventurers' fate are not the same. It differs across age group, whether the tourists wish to avail of the services or not. As is more convenient in the use of technology, the younger generation can carry high-end gadgets with them. Therefore, they might be reluctant to avail themselves of the services of a photographer. On the other hand, the older generations traveling with families wish to capture memories and get them recorded. They might prefer hiring a photographer. Similarly, the services of an adventurer will be required only if one does adventure sports (Table 6.6).

The data suggests that the photographer renders their services mostly to the age group 25 years and above. This age group travels with their family and therefore wish to capture every worthy moment. However, for adventurers, the case is quite the opposite. The services of adventurers are mostly availed by the young age group tourists, i.e., 25 years and below. These tourists engage in adventure activities and therefore require assistance.

6.2.2.17 Average Earnings from Single Tourist

The pricing of photographers and adventurers is bundled together with other things; for photographers, it includes the cost of clicking photographs and developing them in hard copies and packaging them. Sometimes, it includes the cost of delivering the products by sending them using postal means. For adventurers, pricing includes the cost of safety equipment that is needed for performing the sports. It also includes the cost of lodging and food at such destinations if the activity will be performed for more than 1 day (Table 6.7).

The average earning from one tourist for a photographer ranges mostly in a segment of below 1500. The income is on the lower side. Since the pricing has to be kept competitive, therefore, the earnings plummet. In addition, the use of gadgets by tourists has forced the photographers to keep their pricing as low as possible. On the other hand, adventurers earn more in a segment of more than Rs 1000 per tourist. This is because their pricing includes the cost of equipment and necessary items, which increases their overall cost.

6.2.2.18 Technological Usage in Services Provided

Digitalization is apparent in all spectrums of the tourism and hospitality industry. The role of digitalization is large in the case of photographers. Technological advancement in recent years has evolved the equipment. It expedites the process of clicking and developing photographs to just a few seconds. Technology is also used in delivering the photographs. By providing the photos in electronic format, photographers provide an opportunity to get these developed at any place as per tourists' convenience. However, a price is charged for the same. But in the case of adventure, technology doesn't play a very remarkable job. The adventure can be demonstrated to show the difficulties and process that one needs to take care of. A list of things required and database management can be made effectively through the use of technology. However, accepting digital payment is common between both these stakeholders (Table 6.8).

6.2.2.19 Payment Acceptance Via Digital Means

The data suggests that around 82% of the photographers and 71% of the adventurers accept payment in digital mode. However, on digging deeper and enquiring about how much people pay digitally, the data suggests that 75% of the tourists pay in digital means to the photographers and adventurers. This indicates that people are willing to use digital mechanisms if they are provided the opportunity. In addition, since they have to accept digital money, they have an active bank account and other financial services.

6.2.2.20 Feedback Collection from Tourists

Feedback from customers is crucial in the service industry. Suppose the customers are dissatisfied or expect more at the given prices. In that case, the service provider has to respect the expectations. The efforts shall be made to improve the service quality and standards or lower the price to a reasonable measure. Feedbacks can be accepted by various ways. It can be in the form of writing in visitor books, getting

Table 6.8 Accepting digital payments

Response	Photographers	Adventurers
Yes	18 (82%)	15 (71%)
No	04 (18%)	06 (29%)
Total	**22**	**21**

Source: Primary survey

feedback forms filled on the spot, and through social media pages. People are hesitant to highlight drawbacks in services instantly, so feedback through social media is a more reliable form of feedback as customers can review without worrying about facing the service provider.

The data suggest that all the photographers and adventurers collect feedback from the tourists either through visitor book, by getting filled feedback form, or by asking them to review the social media pages' services. These stakeholders also work upon the suggestions. One photographer revealed that it was the request and recommendation of a customer to provide e-copies of the photographs to be easily uploaded on social media accounts. From that point, a separate package was developed that included the cost of supplying e-copies.

6.2.2.21 Services Availed by Tourists in Rural Himachal

The data suggest that in rural Himachal Pradesh, services of photographers and adventurers are not availed. The reason cited for the same is the lack of tourism opportunities in rural areas. It should be noted here that this question is asked exclusively for rural areas. If a tourist visits rural areas after visiting urban areas, the photographer and adventurer accompany them.

6.2.2.22 Problems Faced by Tourism Industry

Photographer and adventurer require a certain degree of professionalism. For this, special training is even required at times. The major problem highlighted by these stakeholders is the lack of professionalism exhibited by certain people. Owning a camera does not make you a photographer. People with little or no experience in this field attract tourists due to their low pricing. Since they do not have the required expertise, they do not entirely live up to tourists' expectations. So, the image of the industry gets deteriorated. Similarly, accidents in adventure sports due to unprofessional behavior have tarnished the industry image. The seasonality of the industry is also a troublesome aspect for these stakeholders.

6.2.2.23 Views on Promotion and Development of Rural Tourism in Himachal Pradesh

The success of rural tourism development and its promotion is perceived thoroughly throughout our survey. Every stakeholder is particular about its development and success because of the presence of rurality throughout the state. The photographers also suggest that people these days wish to explore something new to be displayed as something novel and unique to gain social media following. Rural Himachal Pradesh, as is unexplored, could serve the above-said purpose beautifully. Adventure tourism can also be a hit in rural areas as individual traditional sports and games can be played to attract tourists that promote the state's culture. Both of these stakeholders are hopeful for the industry's development as it will lead to economic gains for them.

6.2.2.24 Availability of Services in Rural Himachal Pradesh

Both photographers and adventurers refused to provide service in rural areas in current scenarios owing to less demand. However, if rural tourism is developed, these stakeholders are willing to perform their rural areas. An increase in the number of tourists will improve their condition as well. They also suggest that seasonality in the tourism industry can also be reduced to a large extent by developing rural tourism. In the lean season of urban areas, tourism can be promoted in rural areas. However, that will be highly challenging as in the rainy season establishing connectivity with the rural areas is an uphill task.

6.2.2.25 Opinion on Future Rates/Charges

Increasing competition and resultant price wars have milked out the profits from the industry. Competitive pricing has reduced price to a very reasonable level. Therefore, the pricing charged by the photographers will not be much different from the prices charged now. In addition, since the product is not unique to urban areas, there will not be any quality issues to the product offered in rural areas. The main typology of tourists visiting the rural areas will be the honeymoon couples or trekkers, who will require good-quality photographs to relive their memories. It will be a one-time memorable experience for them. Adventurers reveal an increase in prices to be charged in rural areas. Since the demand will be initially low, the cost of equipment will be high per tourist. Thus, at first, the costs will be higher, and as the demand advances, the costs will be brought down.

6.2.2.26 To Explore Possible Hindrances to Rural Tourism in Himachal Pradesh

Ensuring rural areas' connectivity and accessibility throughout the year will be a significant obstacle for the rural tourism industry. As landslides are common in the rainy season, many areas get disconnected from the cities. This will harm the tourism industry in such areas. Further, maintaining infrastructure quality will also be an issue highlighted by them. The availability of raw materials and electronic goods has to be necessarily maintained. These stakeholders also lament the inferior branding of the tourism products by the planning agencies.

Since not much technological infrastructure exists in rural areas of the state, if a person wishes to capture photographs, it will depend on photographers. Additionally, by limiting the gadgets that a person can bring, photographers' requirements can be enhanced drastically. Suppose adventure activities are part of rural tourism programs, such as traditional sports or fishing, or similar activities. In that case, there will be a requirement of adventurers. For that purpose, training must be given to the people specifically for that adventure. Since there is not much business opportunity, photographers and adventurers are reluctant to provide their services in rural areas. But they are hopeful that as the industry will progress and develop in rural areas, the demand will increase substantially. Then they are willing to work in such areas. There will be no quality and pricing issues as the pricing is not based on the operation area rather than the quality of work intended. As the same quality and professionalism will be required in operation, the pricing will be marginally lower for photographers and marginally higher for adventurers. These stakeholders' main expectation is the availability of goods required to perform their operation in the nearby market. In addition, the state should ensure that only professionals are allowed to work on their field in the industry. A state certification should be made compulsory for this purpose. As electricity is necessary for their work, photographers require that a continuous supply of electricity at reasonable prices should be there in rural areas to not act as a hindrance in their operation. However, both of these stakeholders are positive about promoting and developing rural tourism in Himachal Pradesh.

6.2.3 Significance/Observed Role of Travel Agents/Tour Guides

Other vital stakeholders who have significant power over the tourists in altering their behavior are the travel agents, tour operators, and tour guides. By either planning trips for them or influencing their behavior in terms of consumption patterns, these stakeholders can result in severe economic gains for the industry. The significance of tour guides rises exponentially if there are more destinations to cover in an area with a limited time available and destinations differ hugely in typology. They also impact the boarding lodging and food pattern of the tourists. What to consume, where to consume, where to stay, what to see, and how much to see all depend upon these stakeholders. They also impart knowledge about the social and cultural background of the destination. These are responsible for the social inclusion of the state resulting from tourism. Because of such high influencing abilities, the perception of these stakeholders for rural tourism development and various tourist behavior aspects must be assessed.

6.2.3.1 Options Availability for Alternate Occupation

As tourism is a season-based industry, the stakeholders need to undertake an alternate employment work during the lean season. Many of the stakeholders associated used to work in some other industry before joining it. Some even continue to work in that industry and are part-timely associated with the tourism industry for additional revenue. Most tour guides worked in factories or as a cook in restaurants in their previous employment and currently pursue the same during their offseason. Some of the tour guides have professional knowledge and had pursued an education in tourism subjects before joining the industry.

6.2.3.2 Available Marketing Options for Customers

To exercise power, one must create an impression of having power. Referential power comes from an association with people who are big shots in their industry. Marketing by tour guides is done similarly. By attaching themselves with big B2B travel companies, they create a market with themselves. Others have their social media pages where they frequently post videos and interact with prospective clients. Few tour guides have also tied up with hotels and restaurant owners for promotion. Travel agents frequently advertise their packages on social media and print media as well. They have their designated website where they frequently post reviews of their satisfied customers for other people to see. Mails are sent to offices where going on outstation picnic and conferences is common.

6.2.3.3 Age Group of Customers Catered

The age group of tourists availing services of travel agents and tour guides does not differ drastically. Since young people are more gadgets friendly and spend their time on social media and other Internet platforms, they do well research to plan journey. They have pre-read about the famous destinations and have high knowledge about destinations and their surroundings. Therefore, tour guides are usually hired by the older generation of people. Services of travel agencies are also available by the generation in the age groups that cannot make an unplanned journey. This age group is also the older generation. In addition, young couples who are on their honeymoon get their tour planned by the travel agents to face no issue during their journey. The table below lists the age group of the tourists last catered by the respondents (Table 6.9).

On the basis of tourists served, the data suggests that tourists consuming cultural or historical tourism are the ones who avail the services of tour guides. They are searching for information that is significant for their research and is not available on the Internet in authentic form. International tourists are the significant customers of tour guides in Himachal Pradesh. Travel agents mainly serve couples who wish to spend their time in solitude and prefer hassle-free travels. Other clients include families going on vacation in hill stations and those going in peak season and want to ensure hotel bookings (Table 6.10).

6.2.3.4 Average Earnings from a Single Tourist

Earnings of travel agents depend more on packages offered by them. These packages differ drastically in their pricing depending upon the typology of tourism being practiced and the number of people visiting. For this reason, it is not realistically possible to estimate the earnings of travel agents from one person. However, the respondents answered that the minimum package they have starts in the range of Rs 5000–5500 per person for a 3-day trip to a particular district of the state.

Table 6.9 Age group of customers served by tour guides and travel agents

Age group	Travel agents	Tour guides
Below 18 years	00	00
18–25 years	03 (5%)	05 (13%)
25–35 years	10 (18%)	12 (32%)
35–45 years	28 (50%)	16 (42%)
More than 45 years	15 (27%)	05 (13%)
Total	**56**	**38**

Source: Primary survey

Table 6.10 Earnings of tour guides from a single tourist

Earnings (in Rs)	Tour guide frequency
1–500	00
500–1000	05 (21%)
1000–1500	08 (39%)
1500–2000	07 (34%)
More than 2000	01 (5%)
Total	**36**

Source: Primary survey

Table 6.11 Use of technology in providing services

Response	Travel agents	Tour guides
Yes	56 (100%)	34 (90%)
No	00	04 (10%)
Total	**56**	**38**

Source: Primary survey

6.2.3.5 Payment Acceptance Through Digital Means

Accepting digital payment is usually standard these days. After the demonetization by India's government in the year 2016, it became really essential to accept digital payments. The practice is in practice to date. It also helps keep money safe during touring to destinations where there is no such safe place to keep belongings. The data suggests that 100% of the travel agents and tour guides accept payment digitally. Tour guides use digital wallet applications such as PhonePe, Google Pay, and Paytm. In contrast, travel agents accept payments from credit cards, debit cards, and even cheques along with mobile-based payments. Altogether, both the stakeholders offer a plethora of payment options to the tourists.

6.2.3.6 Technology Usage in Offered Services

The use of technology is inevitable in the hospitality industry. Different stakeholders use technology in forms that suit their operations and increase efficiency and effectiveness (Table 6.11).

As far as travel agents are concerned, all the respondents agreed that they use technology to provide their services. They detail their packages to the tourists with the help of a video or 3D visualization. The details about the weather are also told to the customers in advance, along with constant reminders about their tour itinerary on mails and messages. Grievances are handled on online platforms, and any required documentation is done using the electronic mode. Tour guides, on the other hand, make use of technology in bringing efficiency to their work. They learn to speak or translate languages for foreign tourists, use imagery software to detail any historical figure or artefacts using the network, tell exciting stories and folklore about places listed over the Internet where details are missing, and accept feedback digitally.

6.2.3.7 Feedback Collection from Tourists

Constructive criticism is the key to progress in life. The key to surviving in the industry is understanding and accepting customers disliking or failure to meet customer expectations and work on it. More than 90% of the tour guides directly ask customers to provide feedback on their liking and disliking to deliver services. This is either done through face to face interaction or with a visitor book or feedback form at the end of the tour. Few guides also make use of their social media pages to collect tourist feedback. Travel agents post their customer reviews on their social media pages for publicity and promotion. For this, they get a testimonial filled from the tourists after they return from their trip.

6.2.3.8 Recommendation Offered for any Place or Dish to a Tourist

Tourists depend on tour guides for a lot of information. Enquiring about the local traditional food and the shops preparing authentic food is one of them. It is also the tour guides' responsibility to recommend tourists food dishes that are local and unique to the state. This way, food tourism can be promoted without any additional cost. Tour guides have recommended dishes like Siddu, maal pude, chilta, Dham, and madra. They recommended them the dishes, but tour guides also directed them to the shops and eateries where they can get authentically prepared Himachali food. Travel agents have also started including food tourism in their tour packages by offering at least one signature Himachali dish to the tourists during their stay.

6.2.3.9 Problems Faces in Tourism Industry

The main problem highlighted by the tour guides is that the itinerary developed by the travel companies is very common. Still, the customer wants some more experience-based products relating to the local community. Thus, tourists look for other states at times where a lot of focus is given on connecting tourists with local culture and traditions. They also lament the minor role of the government in supporting them during the offseason. The advent of foreign companies in the travel agency business has also impacted the industry's economic gains. An interesting takeaway from one of the respondents is that the state does not cater to high-class tourists even though it can do so. The elite class, therefore, does not include the state in their vacation list. Focus on mass tourism has also deteriorated the attraction of the state for the elite class. Due to over-tourism, service standards are falling quite drastically. There is a lack of professional training in the hotel staff, photographers, and tour guides. The same problem persists. The policies for the government for sustainable development have not been able to achieve their targets. Accessibility issues remain persistent in the industry throughout the state.

6.2.3.10 Views on Development of Rural Tourism in Himachal Pradesh

Tour guides and travel agents share similar opinions about the development and promotion of rural tourism in Himachal Pradesh. Tour guides agreed to the possibility of developing and promoting rural tourism in Himachal Pradesh as there is the availability of vast rural areas with local customs and traditions to support the industry. Further, they revealed that explorers, international tourists, and culture enthusiasts look forward to such destinations during their visit. Ninety-four percent of the travel agents too agreed with the proposal. However, they were doubtful over its sustainability as arranging tourists for rural tourism will be a challenge for the industry. They believe that the number of rural tourists will not be such that it will be sufficient to maintain the infrastructure. In addition, including such destinations in existing packages will lead to disturbance in their current pricing policy. In the absence of the required infrastructure, it is deemed impossible to divert tourists to rural areas.

6.2.3.11 Outlook on Services Provided if Given a Chance in Rural Himachal Pradesh

The tour guides and travel agents are willing to work in the rural areas if rural tourism is developed, and a consistent demand arises. Tour guides' scope will be limitless if rural tourism is developed as tourists would want to know about the stories behind local customs, folklore, and traditions. Since not much information will be available over the Internet, the demand for tour guides will surge at such places. The need for travel agents and their associates will arise to make tourists stay more convenient and hassle-free.

6.2.3.12 Possible Changes in Prices/Rates/Charges

The pricing of tour guides will be higher than the currently charged. This will be because of much more dependency of tourists on tour guides. In the absence of more supply and a flourishing demand, tour guides can raise their prices reasonably. In addition, it will include the cost incurred in acquiring authentic knowledge about rural areas. The packages designed by travel agencies will include other costs as well. Therefore, the package's cost will depend on several other factors and hence cannot be predicted. The cost for tourists is sure to increase as adjustments for accommodation, traveling, and food will have to be made in surroundings with limited infrastructure ensuring tourists safety. As the industry will flourish and infrastructure gets better, the prices will reduce to their normal.

6.2.3.13 Major Hindrances to Rural Tourism in Himachal Pradesh

As suggested by the stakeholder, the main hindrance to rural tourism is the lack of marketing in this regard. It will be difficult for policymakers to adjust rural tourism in their existing marketing plan, which is also ineffective in serving the objectives. Further, people's changing mindset where they are getting more inclined to western culture and forgetting their grass root culture will also be a challenge to overcome. There is a chance that rural tourism will also be considered as a sub-standard tourism typology because of its rural environment where it will operate. This might cause failure for the typology as the negative branding and imaging will harm the industry. The lack of government support to stakeholders and industry might also be a hindrance as per the respondents.

The above analysis of data about tour guides and travel agents will have severe rural tourism development implications. First, as suggested by travel agencies, it will be difficult for the industry to have a consistent inflow of rural tourists. For this, a strong marketing program will be required that focuses solely on rural tourism and acknowledge it as a standard product under the umbrella of the Himachal Pradesh tourism industry. Second, the demand for tour guides will be more in rural areas to make tourists aware of the state's culture and traditions. Thus, it can be concluded that a lot of information can be exchanged with the tourists about the state and vice versa with the help of a tour guide through rural tourism development. This will help in achieving the objective of social inclusion where the state lacks behind. Third, since the demand for tour guides will be more, there is also scope for increasing earnings to this segment. This can lead to substantial economic benefits for the industry. Overall, the development of rural tourism is positive for the industry and tour guides and travel agents.

6.2.4 Availability of Taxi/Cab and Other Transport Services

A destination is worthless if it is not accessible. If tourists feel discomfort in accessing it, then the first thought to cross their mind is to abandon the idea of visiting the place itself. Further, using public transport in the absence of knowledge of the local area is an uphill task. For this reason, taxi, cab, and auto operators share an essential

Table 6.12 Responses of taxi/cab owners on different questions

Particulars	Responses
Own a vehicle	Yes ($n = 37$)
	No ($n = 13$)
Accept bookings online	Yes ($n = 6$)
	No ($n = 44$)
Which cab aggregator have you tied up with	Ola ($n = 6$)
	Uber ($n = 0$)
Average earnings per trip	0–200 ($n = 6$)
	200–500 ($n = 10$)
	500–800 ($n = 30$)
	More than 800 ($n = 4$)
Average customers per day	0–5 ($n = 21$)
	5–10 ($n = 17$)
	10–15 ($n = 8$)
	More than 15 ($n = 4$)
Recommended destinations to tourists	Yes ($n = 44$)
	No ($n = 6$)
Drove tourists to recommended destinations	Yes ($n = 35$)
	No ($n = 9$)
Can rural tourism be developed?	Yes ($n = 47$)
	No ($n = 3$)
Will you serve in rural areas?	Yes ($n = 42$)
	No ($n = 8$)

Source: Primary survey

role in the hospitality industry's development. By taking tourists to places at an economical rate safely, these stakeholders help bridge the distance between the demand and supply sides. The tourism industry will not be able to thrive in the absence of these people. Therefore, while developing the rural tourism industry, their perception is also important (Table 6.12).

6.2.4.1 Vehicle Ownership

Taxi, cab, or auto drivers usually follow either of the two mechanisms. First, they either drive their vehicle, or second, they drive a rented vehicle. In case a person drives their vehicle, the chances of earnings are more as it reduces the expenditure on hiring a driver. In a rented vehicle, a person pays a commission on the vehicle to the real owner. Our survey revealed that 74% of the respondents drive their vehicle and others drove a rented vehicle. The main advantage of its vehicle is the absence of dependency on drivers and certainty of earnings because of mistrust by the parties involved. Likewise, those driving leased vehicles do plan to buy their vehicle in the coming years.

6.2.4.2 Acceptance of Online Cab Aggregator's Platforms Such as OLA and Uber

Cab aggregator platforms such as Ola and Uber have mushroomed up in tier 2 cities in India. On the one hand, they offer the public to travel at a very affordable rate by offering discounts and cab sharing features hassle-free; on the other hand, they reduce the need for cab owners' marketing to attract customers. The platform assists both the tourists and cab owners to bridge the gap between them. Thus, cab owners are willing to list themselves on these platforms and pay a certain commission per trip in exchange.

The survey suggests that not many taxi and cab owners are keen on utilizing the platforms available. The main reason cited by the cab owners is the high commission they need to pay to these cab aggregators. Further, the pricing charged is standardized, whereas, presently, the cab owners try to milk as much as possible from the tourists. So, the fall of earnings because of such cab aggregators have forced the cab owners to not collaborate with such platforms to a much extent. There are very few cars running that are associated with these platforms.

6.2.4.3 Other Means to Attract Customers

To showcase themselves to the tourists, these stakeholders follow traditional methods. They have tied up with a hotel, homestays, or a guesthouse and are available for the tourists on a call. In addition, they work in informal collaboration with other cab owners. They move in clusters and do not engage in heavy price wars among them. Specific standards have been set in pricing for trips from famous destinations followed religiously by all the cab drivers.

6.2.4.4 Average Amount Paid by Customers

The taxi driver's economic status is more dependent on if the vehicle is leased or owned by them. In addition, the trips charged by them depend on if the tourist's avail of a package designed that covers different tourist destinations or is a journey to a single destination. Many tourists hire a particular taxi to show them different places that help them in hassle-free travel.

The data suggests that the estimated earnings from customers per trip are more in the range of Rs 500–800 (60%). This is because, in this price range, the tourists quickly cover three to four destinations. Suppose one visit in Dharamshala, then the price is charged for visiting famous destinations like Dharamshala stadium. In that case, Bhagsu falls, the residing place of Dalai Lama and Tea gardens, is Rs 800 for a private taxi. Similar rates are charged throughout the state contingent on the number of destinations covered. Hiring one taxi for their journey is beneficial for both the tourists and the taxi driver.

6.2.4.5 Number of Customers Catered in a Day

The total earnings per day additionally rely upon the number of tourists catered in a single day. It is the average customer bill product and the number of customers catered to a single day. A cab owner can maximize its earning by increasing the number of tourists served in a day or increasing the tourists' average bill. The more preferred way is by increasing the bill charged to a customer. This is because it reduces the need to capture tourists again and reduces the idle time in operation.

The data collect suggests that around 42% of the taxi drivers serve zero to five tourists per day. This is consistent with the principle of increasing the average bill and compromising on the number of tourists in a day. Engaging with one tourist for a long time is beneficial for these stakeholders as their idle time is reduced considerably.

6.2.4.6 Alternative Source of Earnings in Seasons with Low Tourists

At the point when the season is on, the industry is running high. However, as the lean season arises, the revenues from the industry dry up. It influences more to those whose earnings solely depend on tourist arrival. Increasing competition makes it difficult to survive in the lean season. This forces the stakeholders to the alternate sources of revenue. Alternate sources of employment include farming, horticulture, and cattle rearing. Those who do not have access to such farming activities engage in working in shops, factories, or work in specific private organizations. Despite the fact that the earnings are minimal, they are sufficient to wait for the peak season. Many people also engage in renting out their cab during the lean season for other works.

6.2.4.7 Recommendations on Tourist's Destinations

Many tourists engage themselves in discussions with the taxi/cab drivers to know more about the places. They act as sources of information which is not easily accessible or accessible but are not transparent enough. It includes recommendations about destinations, places to stay, food to consume, and particular Do's and Don'ts during their journey. If information is proven to be authentic even once, it helps in building up trust between them. Specific destinations negotiate with the taxi/cab drivers to bring tourists to them.

Mere recommendation of destinations is not enough for the industry. What is essential is that tourists change their planned journey and visit a destination. This way, the referential power of these stakeholders can be determined. It is also much easier for them to take tourists to the destinations recommended themselves. However, it is a risky venture, as well. If tourists are unsatisfied after visiting the recommended destination, they will not much heed to the future recommendation. There is a chance of spoiling the customer satisfaction level as well. For those who did not recommend destinations, it was either because they were hired for very short purposes or the tourists did not intend to visit other destinations and therefore did not engage themselves in any such communication.

Around 80% of the respondents traveled to the recommended destinations with the tourists. While 20% did not do so. The reason cited for not traveling to the recommended destinations is that tourists had their journey pre-planned and did not wish to change because of their large tourist size. However, there are those who could generously accepted suggestions. The satisfied tourists themselves asked for future recommendations about destinations and other related things.

6.2.4.8 Problem Faced in Business

There are problems associated with every stakeholder in the tourism industry. Highlighting these problems is crucial as enhancing the stakeholders' wealth and removing their grievances should be the purpose of study. The referential power of these stakeholders over tourists is significant. Therefore, unhappy stakeholders will never serve the industry effectively and efficiently. The main problems highlighted in the survey include increasing competition among taxi/cab drivers and the threat of cab aggregators carving out the profit share, the problem of employment during the lean season, traffic problems due to overcrowding of vehicles during peak season, poor condition of roads leading to increase in the cost of maintenance of the vehicle, and the increasing fuel prices causing higher operational cost. The increasing bargaining by the tourists due to standardized and low fare pricing by cab aggregators is also hampering the transportation industry.

6.2.4.9 To Overcome Language Barriers with International Tourists

To overcome linguistic barriers with national tourists from non-Hindi- and non-English-speaking regions and international tourists, taxi/cab and auto drivers use technology. With the use of the Internet, they can search the destination's correct address as per the tourists' directions. Further, with the help of translating applications in their smartphones, they can communicate with tourists in their native language. As the tourists from various international countries are frequent, these stakeholders have developed a basic idea about the places that such tourists will be willing to travel to.

6.2.4.10 Views on Rural Tourism Promotion and Development in Himachal Pradesh

From the above discussion, it is clear that taxi, cab, and auto operators are essential stakeholders in the hospitality industry. Therefore, their perception of the promotion and development of rural tourism must be assessed. This analysis can assist the policymakers as these will include grievances addressed while developing the product.

The responses from these stakeholders reveal that rural tourism development in Himachal Pradesh is conceivable. They likewise consent to rural tourism's success because of the presence of vast rural regions in the state. Few respondents also revealed that the main attraction for tourists is the authenticity and traditionality of things. Anything that has local value or attachment appeals to tourists more than global things. This is because it creates uniqueness in the list of items they own. As rural tourism encompasses both these components, rural tourism can turn into a hit tourism typology.

6.2.4.11 To Seek Services If Given a Chance in Rural Himachal Pradesh

Transportation service is a significant part of the tourism industry. So, if rural tourism is to be developed, the presence of transport services will be essential. These services make a destination accessible, and accessibility is a significant tourist attraction. So, these stakeholders' minds should be assessed on whether they are willing to serve in the rural areas of the state or not.

Around 84% of the respondents agreed to provide their services in rural areas if an opportunity emerges due to new tourism products. The main reason cited for this is the opportunity to earn more by increasing the customer base. This will reduce the increasing burden of competition in the industry. However, these respondents' only demand was the improvement of the condition of roads in rural regions and some guarantee of tourist arrival in the initial years. The main reason for the reluctance of respondents to work in rural areas is the roads' dangerous condition that will make their costs higher.

A key observation from the discussion is that the pricing will be different from the current pricing system. Since in rural regions destinations will be much closer, the pricing will have to be regulated. An estimated 20% fall in the pricing was told by the respondents. But, there are chances that the pricing might increase if the running costs are found to be higher than current.

6.2.4.12 Major Hindrances to Rural Tourism in Himachal Pradesh

On discussing the main hindrance—they feel will be in developing rural tourism—the respondents cited that maintaining accessibility and rural infrastructure in such regions will be a challenge for the policymakers. Further, since there will be no local demand for rural tourism, only if consistency is ensured in tourist arrival will only rural tourism survive.

From the discussion with the taxi/cab and auto owners, the following essential implications are derived for rural tourism's effective development. First, these stakeholders are willing to work only if tourists' consistency is ensured in rural areas. Certain improvement is made in rural infrastructure. Second, the pricing will be lower or higher than the current pricing based on operational costs in rural areas and demand. In case the pricing is higher, then necessary measures will have to be taken to remain price competitive. Third, certain regularization in pricing is a must as customers might feel cheated in its absence. Fourth, destinations should be available as a package. This is because these stakeholders prefer to stick to one tourist for a long journey. This additionally fabricates trust between the parties and develops referential power on the stakeholder's end. And last, the accessibility between destinations in rural regions must be improved. Better conditions of roads and communication networks will be crucial for the survival of the transport industry in rural areas.

6.2.5 Inclusion of People Representatives/HPTDC Officials

There is another segment that is actively associated with the tourism industry. This include Himachal Pradesh Tourism Development Corporation officials that work actively for the development of schemes and policies for the tourism industry and people representatives who are in touch with the local people and put forward issues related to industries at the regional or national level and academicians who are engaged in research activities related to the tourism industry. These people know the reality of tourism industry scenarios and trends. The results from the schemes launched by the government are known better to the researchers and policymakers themselves. Therefore, understanding their perception of the development of rural tourism is essential.

6.2.5.1 Significant Problems Associated with the Tourism Industry

On the question of challenges faced by the tourism industry in Himachal Pradesh, the respondents highlighted tourism planning, carrying capacity estimation, mass tourism, ignorant tourists, lack of government vision for sustainable tourism development, mass awareness, quality of workforce, and nonprofessional into industry. Diversified products, multiple controlling agencies, proper connectivity, infrastructure, and seasonality are the significant challenges. Ineffective marketing of the tourism industry is one of the major challenges revealed by the respondents. These challenges can be categorized into administration related, infrastructure related, weakness of the tourism industry, and autonomous factors. The challenges that are within the scope of correction are the ones that originate due to the administrative weaknesses or characteristics of the industry. Sustainable development is one of the solutions how the major issues highlighted can be settled. The issues discussed are not specific to Himachal Pradesh. Hence, these can be dealt with in how the states practicing sustainable tourism have tackled them.

6.2.5.2 Different Active Schemes for the Development of the Tourism Industry

The role of government in setting up and developing an industry is significant. The government acts as an incubator for the industry. Through its financial and nonfinancial support, it tries to eliminate the industry's hurdles. The government's primary role highlighted is the tourism promotion by government of India and state government. Apart from these schemes about Himachal Pradesh include heli taxi promotion during festival season, ADB Funded research for the development of the USP, and Home Stay. These schemes focus on developing niche tourism products by offering financial, technical, and marketing assistance. Improvement of accessibility, infrastructure and is also a central component of schemes launched by the government.

6.2.5.3 Attitude of Locals in Region Concerning Tourism

The survey revealed that the local population has a positive outlook for the incoming tourists. The local population is keen on welcoming tourists because of their livelihood dependence. Income in the form of earnings from the hotel, homestay rent, food sale, and shopping by tourists creates flourishing economic conditions in the state. However, respondents criticized the attitude of the local population. They objected to the money milking attitude of locals that is causing harm to the environment.

6.2.5.4 Ways to Deal with Limited Natural Resources Available and the Increasing Inflow of Tourists

The survey revealed that researchers and policymakers' focus is educating the locals about the assessment of carrying capacity and serving the tourists accordingly. Efficient resource management is one of the other ways through which the issue of lack of resources is tackled by the policymakers. A new method of serving tourists on a pro-rata basis is also emerging in the state to tackle resource shortage, particularly in areas that have high tourist footfall and frequently face a lack of assets in peak seasons. The researchers study the gap between carrying capacity and tourist inflow to develop a model that will help serve tourists in sustainably.

6.2.5.5 Dependence of Region's Economic Development on Tourism

The survey reveals that the economy of Himachal Pradesh is very much dependent on the tourism industry. Although horticulture is practiced well within the state, people try to attach themselves with the industry during the peak tourism season in some of the other forms to gain economic benefits (Table 6.13).

The data suggests that in no way Himachal Pradesh can afford to make rules for tourism stricter to achieve sustainable development. This will reduce the economic benefits and thus will harm the economy to a large extent. If possible, there should be alternate employment opportunities in the tourism sector in an organized manner.

Table 6.13 Economic dependence of region on tourism

Response	Frequency
Very low dependence	6 (20%)
Moderate dependence	4 (13%)
Good dependence	6 (20%)
High dependence	14 (47%)
Total	**30**

Source: Primary survey

This will help in maintaining the records of people employed and provide training to them for future development.

6.2.5.6 Significant Problems Identified in Response to Tourism

The tourism industry is not a smokeless industry, i.e., it has caused several social, environmental, and economic imbalances within the state. Putting pressure on the scarce natural resources, disturbing the state's social harmony, harming the culture and values by polluting the mind of youth are some of the ways tourism has released toxins in the state. Stakeholders suggest that garbage, air-water pollution, unplanned urban growth, loss of local culture, the problem of early age earning, overcrowding in peak season, and pollution are the primary damage caused by the tourism industry in the state.

6.2.5.7 Opinion on Flourishment of Rural Tourism Industry in Himachal Pradesh

Of the responses recorded, 90% of the tourists suggested that rural tourism will flourish in the state because of the presence of high rural areas and exciting tourist destinations in rural areas. The stakeholders also cautioned about the use of professionals in rural tourism planning and development. However, few respondents said that rural tourism will not be booming in the Indian context. It provides and results in an ambiance that is not as per the expectation of the Indian tourists.

6.2.5.8 Reasons Concerning No Drastic Step Taken Concerning the Development of Rural Tourism

Respondents cited several reasons for no serious government approach for developing the rural tourism industry in Himachal Pradesh. The prominent reasons include lack of research and development in rural tourism themes, less involvement of the local rural community in developing rural tourism, and lack of expertise in the system. There is no network among state government-academician-tourism professionals. This lack of coordination at the planning and policymaking level has created a rift between the stakeholders. The planning is done. As a result, it is far away from reality.

6.2.5.9 Ways to Advertise and Propagate the Schemes to the General Public as Well as Local People Regarding Tourism

Schemes are a waste if they are not known to the people for whom they are created. Therefore, the marketing of schemes is crucial to reach the public and avail of its benefits. For marketing the schemes, respondents suggest that with the help of the District Tourism Development Office at Reckong Peo, the Block Development Officer's office, and village panchayat and through teaching and awareness, disbursement of brochures, flyers at the tourist information office, door to door and advertising, and the Internet and social media, the schemes and associated benefits are communicated to the people. Those who avail of the benefits of schemes are then asked to pitch the plans in their operation area so that others can also avail of the benefit.

6.2.5.10 Technology Incorporation in the Development of Tourism Destinations

Technology is considered to bring about a significant revolution in the tourism industry. The respondents also acknowledge the impact of technology on the tourism industry. They suggest that social media has been more useful than anything else to promote tourism. However, it has led to mass tourism and great devastation in some regions, e.g., Leh after 3 Idiots. Technology is playing a pivotal role in teaching in the tourism industry. With access to the 3G and 4G networks, the lesser-known destinations are publicized. The hotel is equipped with Wi-Fi, a DG generator, and a solar heating system. These are only technology being used in which the rest of the tech people have no idea, like water in heating management during harsh winters. The use of information technology in database management is prominent. Research in tourism themes has covered leaps and bounds due to the advent of technology. Publishing marketing research and database is possible due to the technology. Branding through online mode is also an emerging trend in the tourism industry. However, respondents suggest full support from the local and central government to fully implement digitalization in the tourism industry.

6.2.5.11 Funding Sources for the Development of Tourism Destinations

Finance is the bloodline of every industry. In the absence of required finance, good investment proposals are left out. For financial support, the tourism industry in Himachal Pradesh is dependent on the government. Grants are given by the government to set up homestays, or another financial assistance provided under its different schemes is the primary source of funding for the stakeholders. Public-private partnership is an emerging trend in meeting the funding requirement. However, only marginal projects are sanctioned through this method. People likewise acquire funds from national and regional banks to meet their requirements.

6.2.5.12 Destinations Exhausted and Need to Discontinue Tourism

If there is any tourism destination that should be discontinued due to overburden and depletion of natural resources, the responses show contrasting results. 50% of the tourists suggest that almost all the famous tourist destinations in the states such as Shimla, Kullu, Dharamshala have breached their carrying capacity. Therefore, the need is to discontinue these destinations. The remaining respondents said that although the situation is serious, but it has not reached a level it can be termed a critical ailment. They suggest restricting tourism to such destinations rather than discontinuing them. The reasons cited for the same include economic benefits associated with the locals from the tourism industry. They comment that in the absence of tourism, the local population will have to search for alternate employment opportunities. This will put pressure on the government to generate employment in such areas.

6.2.5.13 Main Occupation of Rural People

People in Himachal Pradesh are majorly dependent on agriculture, horticulture, and tourism for their livelihood. During the peak tourism season, most of the people get associated with the industry in some form. As the lean season begins, they get themselves engaged in agriculture and horticulture activities. People also get associated with dairy farming and cattle rearing. The individuals who can't perform farming and non-farming activities engage in commercial activities such as general merchant shops and jobs such as laborers and semi-skilled workers in other industries. In the primary tourism season—agriculture in the tourism off-season—those who cannot perform agriculture tend to move out to other regions searching for seasonal jobs. Other occupational activities include sheep rearing and farming, horticulture, agriculture, and rearing animals. Few people are also employed in other private and government jobs such as school and college teachers, working as accountants. Alternate job opportunities, i.e., occupations other than tourism, include agriculture, animal rearing, handicraft, small household industries, and slate mining.

6.2.5.14 Food Dishes Promoted in the State Bhawans

The responses show that state Bhawans located in the different states across the country promote Himachali Dham in their canteen. Dham's demand in their canteens is encouraging, and efforts are made to make people aware of the other dishes through marketing and serve them along with Dham. These dishes include madra, Siddu, momos, and few traditional beverages.

The above discussion has severe implications for the development of rural tourism in Himachal Pradesh. First, marketing and accessibility will prove to be the major challenge while developing rural tourism. In the absence of both or any of them, the desired results will not be achieved, and the industry's performance will suffer. Second, developing tourism, keeping in mind the destinations' carrying capacity, is the expert's critical recommendations. An unsustainable, unplanned, and unprofessional approach to the development will lead to the project's failure. Third, several destinations have surpassed their carrying capacity limit at famous tourism destinations. Consequently, there is either requirement to restrict tourism at such places or completely abandon tourism at these destinations. Fourth, economic contribution to the economy of Himachal Pradesh is significant, so the development of rural tourism will prove to be a catalyst in the development of the rural economy of the state. Fifth, the focus of the policymakers on building a state dish is on Dham. Through it, they are planning to boost the food tourism industry of the state. So, Dham should be served at rural tourism destinations to align with the goals of the policymakers. And sixth, the approach for rural tourism development should be scientific and strategic. Professionals should be engaged in planning and saying that rural people should also be there to plan to motivate them to work in the industry.

6.3 Conclusion

The chapter concludes that other stakeholders believe that Himachal Pradesh has significant potential for developing rural tourism. However, results show a knowledge gap that exists between the tourists and other stakeholders. This gap results from the industry's failure to understand the tourist needs and wants and serve accordingly. As visible in the previous chapter, the infrastructure considered essential by the tourists is considered a luxury by the other stakeholders. Moreover, the tourists' perspective on technology as a fundamental infrastructure is being universally asserted by all stakeholders, ostensibly due to its perceived necessity according to tourist preferences. Stakeholders also did not clearly understand the importance and role of social media platforms in attracting tourists. Analysis reveals specific positives for the industry as well. Results show that stakeholders share a cordial relationship with the tourists. The destinations recommended to tourists were visited by them, recommended food dishes were consumed, and stakeholders also incorporated the tourists' suggestions in their services. Stakeholders did agree to serve in the rural parts of the state if the demand exists. This ensures that there will not be any demand-supply mismatch for the industry in rural areas. However, the pricing will be different and will be framed as per the actual scenarios.

Further, the development of rural tourism will also prove beneficial for the industry's economy and social aspects. By interacting with different strata, rural people will get educated about lifestyle and try to upgrade themselves. Increasing demand for goods will cause the market to flourish with the right quality products raising the living standards in rural regions. In addition, by making tourists aware of the local customs and traditions, the state's objective of social inclusion can be achieved. A little is known about the different tribes and cultures of the state. Knowledge of food dishes with the tourists is also primitive. These can be enhanced effectively and efficiently with the help of rural tourism development. As the hospitality industry is using technology at different levels with ease, there is scope for financial inclusion. By making people more habitual in their use, they can be persuaded to use more financial services. Increasing the use of financial services will lead to a more organized financial system in the state's rural areas. But the development of rural tourism will not be an easy task. A persistent and consistent inflow of tourists will be essential to maintain stakeholders' interest in serving in rural regions. Necessary infrastructure will be needed to set up in such regions. The presence of educated and professional staff will be essential to keep the tourism and hospitality industry intact. Marketing programs will be needed to attract tourists, and integrated marketing efforts with the use of all channels will achieve the desired results. What is needed by every stakeholder is a little support from the government in conducting business. It could be in the form of financial assistance in setting up business or assistance in the lean season. Overall, the stakeholders believe that the rural tourism industry's development will be incredibly productive for them. It will open up new opportunities and result in economic gains for them.

Perspective into a Flexible Strategic Crystal

7

7.1 Integrating the Perspective of Various Stakeholders for Rural Tourism Development

Rural development cannot alone mount an impact on the economy of a nation. Subsequent to analyzing the data collected from various stakeholders, i.e., tourists, and the other stakeholder groups involving hoteliers, it is observed that all the stakeholders have a favorable attitude toward the development of the rural tourism industry in Himachal Pradesh. This observation is based on experts' following strengths, from tourist behavior, and on stakeholders' recommendations. These are discussed in the section below:

- Willingness to consume from tourists: The analysis showed that the tourists were willing to consume rural tourism and its related products. Thus, this checks the essential requirement for a product to be successful.
- Willingness to serve by supply side: The supply side of the industry, i.e., hoteliers and all, is also believed that rural tourism will be a successful tourism product. They also showed willingness to serve in rural areas.
- Support from HP demographics: High literacy rate (both male and female) and sex ratio in rural areas suggest people's positive orientation toward females. Thus, it improves the scope of removing socioeconomic challenges through tourism.
- Support from HP geography: The topography of the state is supportive for the development of the tourism industry.
- Favorable opinion of experts: The industry experts are also of the favorable opinion of the success of the rural tourism industry.

Similarly, several pitfalls are highlighted in analyzing the data together. These concerns will pose challenges for the rural tourism industry in HP. These are discussed in the following section:

S. Kumar, *Sustainable Rural Tourism in Himalayan Foothills*,
https://doi.org/10.1007/978-3-031-40098-8_7

- Safety concerns: Tourists and industry experts have raised safety challenges as the primary concern while developing rural tourism. With rural areas existing in non-easily accessible regions, providing security to tourists, mainly female tourists, will be an uphill task.
- Limited technology: The extent of technology diffusion in rural HP will also be a significant challenge identified by the tourists, stakeholders, and industry experts. Availability of electricity and Internet facility might be a challenge while developing rural tourism.
- Low standard of hospitality: Tourists' demands are increasing, as suggested in the survey. The other stakeholders and experts feel that the rural tourism industry might not serve tourists effectively due to lower hospitality standards. Maintain infrastructure; hygiene will be a challenge for the industry. An unsatisfied tourist might not be willing to revisit the state.
- Less economic contribution: The other stakeholders and experts believe that rural tourism's economic contribution will be lower than the contribution made by the tourists in urban regions. Thus, it will be problematic for the industry to survive with lower economic contribution. The stakeholders might not be much interested in diverting their resources to rural regions as the economic gains would be less.
- Low marketing efforts: The marketing efforts of the HP tourism industry are abysmal. Without the necessary funding, the industry cannot focus on framing and monitoring the media plan. The tourists, stakeholders, and the experts too highlighted the weak marketing campaign of the industry. This challenge will be hard to overcome for the industry.

Various marketing masters have highlighted several factors that are crucial for the success of new products. The table below lists these factors and the overall perceived status of these factors of the industry (Table 7.1).

Of the 20 factors identified for the success, eight factors exist for rural tourism and three factors are at a manageable level, and these factors can improve rural tourism. The main concerns are the six factors that need improvement and three factors that are nonexistent for the industry. The three factors that are nonexistent and pose a significant threat are framing a media plan, advertising and publicity, and investing in media monitoring. As it is the result of the perception of the industry stakeholders, we can conclude that the industry needs to work on improving its marketing. However, the eight identified factors are sufficient for the tourism industry to move ahead with the rural tourism product (Fig. 7.1).

7.2 Porter Diamond Analysis of HP Rural Tourism

The porter diamond analysis has four main components and additional two components. This analysis is very helpful in determining the probable success of the industry. The four components for any industry are the following:

Table 7.1 Perceived status of success factors of Himachal Pradesh rural tourism industry

		Perceived present status			
S. no	Factor for success of rural tourism	Very much present	At manageable level	Needs improvement	Nonexistent
1.	Developing a USP		✓		
2.	Framing a media plan				✓
3.	Identify target audience	✓			
4.	Invest in media monitoring				✓
5.	Timing of launch	✓			
6.	Awareness of stakeholders			✓	
7.	Incentives to early visitors			✓	
8.	Support industry status	✓			
9.	Build vision and mission		✓		
10.	Niche launch			✓	
11.	Advertising and publicity				✓
12.	Personalization and customization			✓	
13.	Prepare future goals		✓		
14.	Readiness of demand	✓			
15.	Consent of supply side	✓			
16.	Industry vision of product	✓			
17.	Support from geography	✓			
18.	Support from demographics	✓			
19.	Safety concerns			✓	
20.	Level of hospitality			✓	
	Total	**08**	**03**	**06**	**03**

Source: Author's compilation

- Factor conditions: These include presence of natural as well as man-made resources for the development and maintenance of the industry. The basic resources include useful natural resources and the availability of unskilled labor. Advanced or "created" resources include specialization and skilled knowledge and expertise, availability of capital, infrastructure, etc.
- Demand conditions: These refer to the demand and the market for the product. The size of the market, the growth rate of the market, etc. are some indicators of the home demand.

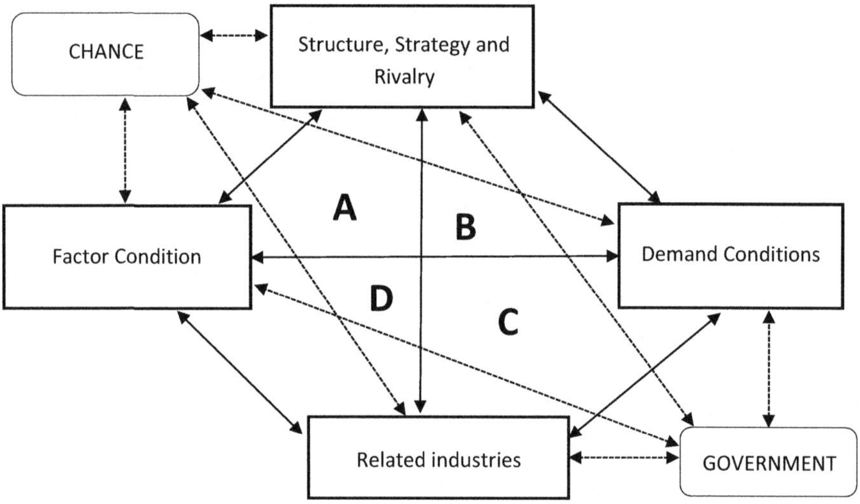

Source: Author's representation

Fig. 7.1 Porter diamond model. (Source: Author's representation)

- Related industries: The level of success of one industry can be related to the success of related and supporting industries. These suppliers help in advancing innovation processes through shared resources: technical and other types of aids.
- Structure, strategy, and rivalry: The internal environment in which a firm is established determines how firms are created and structured. This structuring of the firm can be influenced by a number of factors: political, economic, and social. This structuring will form the basis of creating a strategy toward the establishment of the firm.
- Government: The government plays a vital role in the success of a firm or company. It is the government that provides for technical and financial aid to companies for its growth. The government has been referred to as "a catalyst and challenger."

If one talks about rural tourism in Himachal Pradesh, then it scores good for demand conditions. It is evident from stakeholder attitude that there is demand for rural tourism. Even though the market segment is negligible at present, the potential to grow in such market is high. Then, the factor conditions are present in Himachal for the industry to develop. Natural as well as man-made resources are abundant in HP for rural tourism development. The related hospitality industries in HP are supportive of rural tourism. Even if the industry is budding at present, the positive attitude of the industries will make it possible for industry development. The government is also supportive of the rural tourism development in HP. Thus, the porter diamond is formed for the rural tourism industry in HP.

Since the conclusion is that the rural tourism product will be successful in Himachal Pradesh, the need is to construct a flexible strategic crystal for its smooth and effective development. The following section deals with designing the same. The analysis will be carried out in the following steps:

- Assessment of life cycle stages of tourism products.
- Assessment of life cycle stages of tourism destination of HP.
- Developing the continuity-change (C-C) matrix of the tourism industry.
- Mapping the C-C factors in VDB-HML matrix.
- Developing flexible rural tourism model using interpretive structure modeling.

7.3 Assessment of Life Cycle Stage of Tourist Destinations and Tourism Products

It is said that a product passes through various stages during its life cycle. The life cycle here infers the duration for which a product is available to be sold in a market. There are primarily five stages for a product. The first stage is the introduction stage, where the product is launched in a market, is relatively unknown to the buyers, and does not have much demand. Efforts are made in making customers aware of the availability of the product. The second stage is the growth stage, where the product is getting popular in the market and promotion methods are working well in enhancing the demand for the product. The third stage is the maturity stage, where the product has become well known to the customers and its demand has peaked. There cannot be any increase in demand unless any modification has been made. The fourth stage is the decline stage, where the product starts to lose customers because of its weaknesses or due to increasing competition. Moreover, the fifth and last stage is the relinquishment or relaunching stage, where a decision is taken to either abandon the product or relaunch it with additional features.

This analysis must be carried out not only for tourism products but also for tourist destinations. By assessing these, suitable strategies can be drawn for new products, and decisions can be made with regard to the existing products.

For assessing the life cycle stage, every destination and tourism product is ranked on two parameters. First is the time duration spent in the market since its launch on the X-axis and market demand on the Y-axis. Both the parameters are scored, and the scores are compiled to give the life cycle stage of a product or a destination (Fig. 7.2).

The life cycle assessment of the tourism typology was carried out in the past 70 years of data. For this purpose, research databases were searched for papers and documents that have mentioned the tourism typology. The results show that pilgrimage, leisure, and religious tourism are prevalent in the state since long. Spiritual tourism, historical and cultural tourism, and adventure tourism have passed their introduction stage and are currently in the growth stage. Ecotourism, MICE, rural tourism, wedding tourism, and dark tourism are the new tourism typologies that can grow in HP. This analysis has several implications. First, the number of religious and pilgrimage tourists is reduced for a few years. Also, during the lockdown imposed in 2020, the religious monuments were closed, thus further reducing the

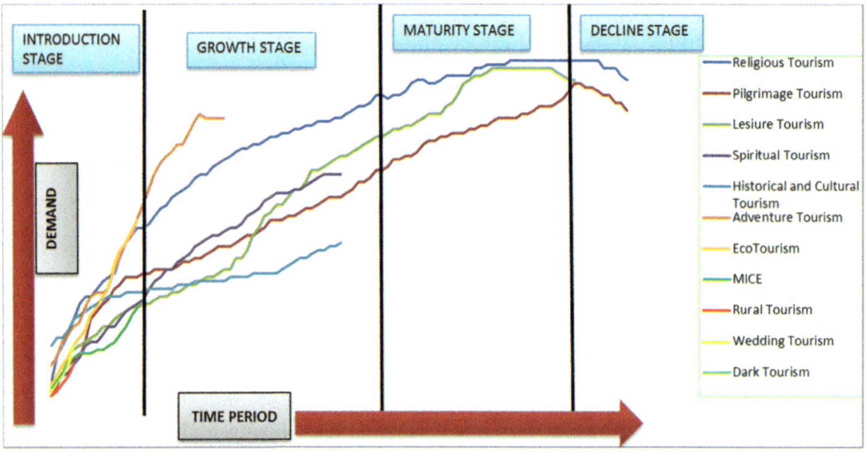

Fig. 7.2 Life cycle assessment of various tourism typologies. (Source: Author's representation)

tourist footfall. The cause of the tourists' decrease must be investigated as it is the oldest typology consumed by the tourists.

On the other hand, the demand for rural tourism, wedding tourism, and ecotourism is increasing over the past few years at an increasing rate. It means that the tourists demand such typologies to be developed and are far more willing to visit HP for such purposes. The policymakers must resort to adopting an appropriate marketing mix for various niches. They can either group niches that are at a similar life cycle and adopt an appropriate strategic market mix or target a niche as per their preference. A BCG matrix analysis of the market will further enhance the knowledge of the policymakers in this context (Fig. 7.3).

On carrying out the life cycle analysis of the various destinations in the state, it is observed that all the possible destinations have reached their specific limits. For this analysis, destinations were grouped based on their districts and the total inflow of tourists was observed for the past 20 years. The data suggests that almost all destinations have exhausted their life cycle. However, some destinations are top rated, and others are not so popular. For example, Kullu and Shimla always show higher tourist inflow within the state. However, in recent few years, the state's tourist inflow is declining in both the districts. This may be due to the environmental concerns arising out of irresponsible resource consumption. This analysis highlights the importance of sustainable tourism development in the state. Some destinations such as in Solan and Kinnaur catch a little interest of the tourists. There may be destinations in these districts that can serve as a tourist hotspot but are unexplored and unknown. A fascinating observation in the data is that in the last few years, the destinations having high tourist inflow suffered more than the destinations where tourist inflow was less. It could mean that the destinations that had less tourist inflow catered to religious and pilgrimage tourists. These tourists visit irrespective of the environmental conditions affecting tourism (Table 7.2).

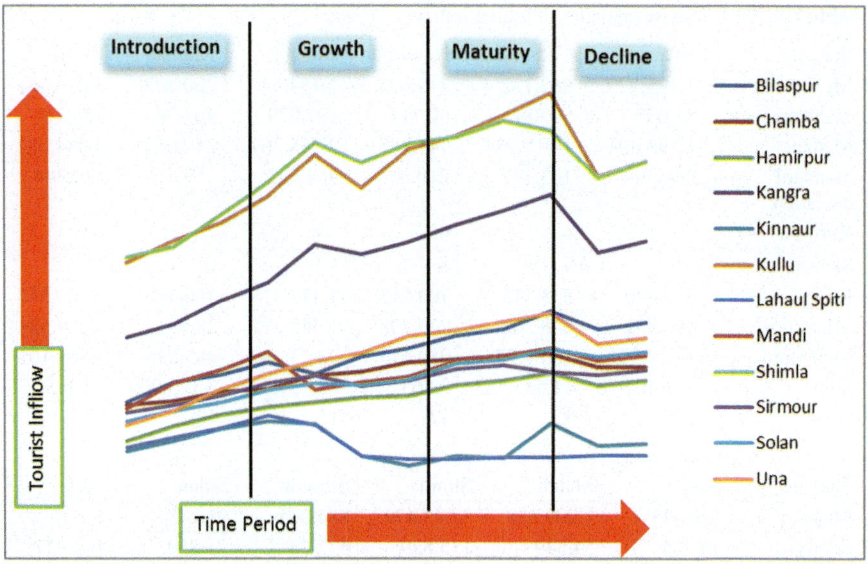

Fig. 7.3 Life cycle assessment of various tourist destinations. (Source: Author's representation)

The descriptive data statistics suggest that the range of tourist inflow is pretty high in Kangra, Kullu, and Shimla. It indicates that the industry has covered leaps and bounds in these districts. The tourist inflow range is lowest in Kinnaur, Lahaul and Spiti, and Sirmour. It indicates that tourists are not that attracted to these destinations. The analysis is significant as it shows the trends of tourist inflow, the highest and the lowest number of tourists in each district over the years. The policymakers must identify the weak performing destinations and run statistical analysis to draw more accurate results.

7.4 Continuity-Change (C-C) Matrix Analysis of Himachal Pradesh Tourism Industry

C-C matrix is a flexible strategy used to determine the "to be" course of action for an entity. Everybody faces a dilemma of whether to continue on its path or explore new opportunities. This dilemma is caused due to two factors – continuity factors and change factors:

- Continuity factors – These are the factors that persuade someone to carry on on the path it is going today. These are the reasons arising out of the entity's strengths or weaknesses to avoid pursuing change.
- Change factors – These occur as opportunities and threats in the environment and are responsible for deviating from the current path to ensure survival and growth.

Table 7.2 District-wise statistics of tourist inflow

Statistic	Bilaspur	Chamba	Hamirpur	Kangra	Kinnaur	Kullu
Mean	1,195,579	962,132	713,633	2,154,527	260,492	3,023,761
Standard error	82,090	52,881	60,711	131,229	46,078	150,210
Median	1,193,404	1,018,934	741,188	2,265,269	240,652	3,084,082
Standard deviation	284,366	183,187	210,309	454,592	159,618	520,344
Kurtosis	−1	−1	0	−1	−1	−1
Skewness	0	−1	−1	−1	0	0
Range	933,420	533,515	700,815	1,477,315	456,465	1,750,517
Minimum	683,951	648,262	279,798	1,345,974	28,581	2,114,584
Maximum	1,617,371	1,181,777	980,613	2,823,289	485,046	3,865,101
Sum	14,346,950	11,545,588	8,563,596	25,854,323	3,125,906	36,285,128
Count	12	12	12	12	12	12

Statistic	Lahaul and Spiti	Mandi	Shimla	Sirmour	Solan	Una
Mean	228,573	1,009,858	3,043,470	884,205	916,481	1,138,789
Standard error	46,374	54,349	133,839	43,974	71,350	104,947
Median	133,068	1,056,806	3,159,926	927,040	897,103	1,237,902
Standard deviation	160,645	188,271	463,631	152,329	247,164	363,547
Kurtosis	−1	0	0	0	−1	0
Skewness	1	−1	−1	−1	0	−1
Range	453,462	629,800	1,407,649	485,593	748,741	1,151,701
Minimum	90,393	612,240	2,174,456	577,583	482,818	446,911
Maximum	543,855	1,242,040	3,582,105	1,063,176	1,231,559	1,598,612
Sum	2,742,872	12,118,291	36,521,637	10,610,455	10,997,774	13,665,469
Count	12	12	12	12	12	12

Source: Author's compilation

Whether an individual, company, organization, or industry, every entity must decide what route it must follow. The tourism industry experiences a similar dilemma in Himachal Pradesh. It has to decide whether to follow the policies and strategies it is following now or make a change if any threat is imminent due to its policies.

The following table shows the continuity and change factors identified from the literature review on the tourism industry (Table 7.3).

A VDB analysis is then carried out on the continuity factors. Vital factors are the most critical factors that act as drivers for an entity to carry on its path. They are of primary importance for formulating plans and strategies. Desirable factors do not have much influence, but their existence provides an edge over others. Burden factors arise due to one's limitations, and these factors must be avoided from policies and strategies. The following table shows the categorization of continuity factors of the Himachal Pradesh tourism industry (Table 7.4).

Table 7.3 List of continuity and change factors

S. no	Continuity forces	Change forces
1.	Existing infrastructure	Threats from substitutes in the existing market
2.	Competitive advantage	Resource availability
3.	Values and culture	Demands of tourists
4.	Market segmentation	Growing focus on sustainability
5.	Current performance	Business ethics
6.	Risk-averse attitude	Changes in internal business environment
7.	Lack of capital	Benefits of diversification
8.	Preferred habitat	Digitalization
9.	–	Niche tourism development

Source: Author's compilation

Table 7.4 Vital-desirable-burden analysis on continuity factors

Category	Factors	Suggested actions
Vital	Infrastructure, value and culture, current performance, market segmentation	Adopt fully; act as suggested by C-C matrix
Desirable	Competitive advantage	Customize to suit mission and vision
Burden	Risk-averse attitude, preferred habitat, lack of capital	Avoid

Source: Author's compilation

Table 7.5 High-medium-low-impact analysis on change factors

Category	Factor	Suggested action
High impact	The demand of tourists, growing focus on sustainability, changes in internal business environment, threat of substitutes, resource availability, niche tourism development	Must consider; act as suggested by the C-C matrix
Medium impact	Business ethics, benefits of diversification, digitalization	Consider the impact but do not fret much
Low impact	–	Ignore

An HML analysis is carried out on change factors. The factors are then classified into high-impact factors, which exert much pressure to change and must be considered while developing plans and policies. Medium-impact factors are those which have considerable impact and cannot be ignored. Low-impact factors do not have high influence and can be ignored. The following table shows the categorization of change factors of Himachal Pradesh tourism industry (Table 7.5.)

After calculating the scores collected from the primary data, the score for continuity forces came out to be 3.21, and for change factors, it was 4.49. On mapping it in the continuity-change matrix, the tourism industry of Himachal Pradesh was found to lie in the synthesizer quadrant (Fig. 7.4).

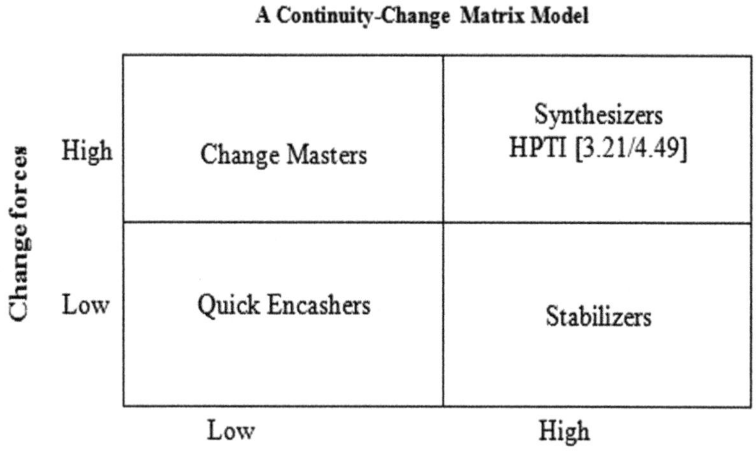

Fig. 7.4 Mapping of Himachal Pradesh tourism industry on C-C matrix. (Source: Author's representation)

As observed, the tourism industry of Himachal Pradesh lies in the quadrant of flowing stream strategy. Flowing stream strategy talks about building up a confluence between continuity and change. One way to achieve this balance is through sustainable tourism development. Niche tourism development is one of the key strategies identified through the continuity-change matrix to gain an advantage for the Himachal Pradesh tourism industry. Therefore, rural tourism development can be termed as a game-changer for the state's tourism industry (Table 7.6).

7.5 Finalizing the Strategic Crystal for Rural Tourism Development

To finalize the strategic crystal for the development of rural tourism, the following process has been followed.

7.5.1 List of Factors Identified from Literature, Tourists, and Other Stakeholders' Perception

To finalize the strategic crystal, we ought to look from the perspective of tourists and other stakeholders. Since the data is available about the tourist and other stakeholders' perceptions, we molded the factors accordingly. So, for analysis, 15 of the factors have been identified for the successful rural tourism development (Table 7.7).

- Unique USP: If the new product being launched has a unique selling proposition or in itself can act as USP of the industry, then the chances of its success increase.

Table 7.6 Strategic framework of continuity and change forces

Continuity factors change forces	Infrastructure	Value and culture	Current performance	Market segmentation	Competitive advantage
Demand of tourists				Red Ocean strategy to increase market share	Gain first-mover advantage
Focus on sustainability		Incorporate environmental protection in vision	Build eco-friendly mechanism		
Resource availability			Innovate to increase efficiency	Cater to new and multiple market segments *(niche tourism development)*	
Threat of substitute					Make competition irrelevant; blue ocean *(niche tourism development)*
Changes in the internal environment		Concerns for the needs of all stakeholders			
Business ethics		Industry culture should be ethical			
Benefits of diversification	Diversify using piggybacking			Increase customer base and market share	
Digitalization			Boost productivity and efficiency		Technology management

Source: Author's own

As the HP tourism industry does not have any USP till date, rural tourism can act as the USP of the industry (Kumar et al. 2018b).

- Marketing of typology: Marketing of the product can itself act as a factor crucial for its success. Rural tourism has to be marketed well to reach public needs. There should be a fund for market plan management as well (Kumar et al. 2019).

Table 7.7 List of factors influencing rural tourism development

S. no	Factor name	Factor code
1.	Unique USP	01
2.	Marketing of typology	02
3.	Willingness of tourists to come	03
4.	Willingness of supply side to serve	04
5.	Infrastructure development	05
6.	Focus on sustainability	06
7.	Resource availability	07
8.	Destination tourism potential	08
9.	Demographics of HP	09
10.	Economic incentives	10
11.	Market segmentation for typology	11
12.	Value and culture of HP	12
13.	Digitalization in industry	13
14.	Current performance	14
15.	Competitive advantage of industry	15

Source: Author's elaboration

- Willingness of tourists to come: A product is successful only if it is demanded. As the data collected has suggested that tourists are willing to come to rural areas for tourism, this confirms market demand for rural tourism (Lal et al. 2019).
- Willingness of supply side to serve: There has to be a balance between the product demand and supply side. It leads to equilibrium pricing. The data collected suggests that the supply side is also willing to render service in the rural areas (Chemli et al. 2022).
- Infrastructure development: Infrastructure serves as the essential requirement for developing any tourism (Kachniewska 2015). Infrastructure such as transportation, lodging, boarding, sanitation, and food is required in rural areas to serve the tourists. It not only aids the tourism industry but also boosts the standard of living in the surrounding region. Infrastructure development leads to an increase in tourist inflow.
- Focus on sustainability: Increase in environmentalism has influenced the tourism industry a lot (Fernández-Hernández et al. 2016; Osti and Cicero 2018). Now, people have become conscious of the ill influence of visiting a place. There is an economic loss to the industry because of the reduction in tourists' number due to the decreasing popularity of mass tourism products. By offering rural tourism as an alternate and extended tourism product, many tourists could be attracted to it.
- Resource availability: Resources are the means to carry out the operations. For tourism activity, resources can be financial, human, and natural. All these resources are available in HP, although in limited quantities. So, effective resource management will pose a challenge for the industry (Kumar et al. 2018a).
- Destination tourism potential: Destination characteristics influence the tourism potential of the destination. Thus, the destinations needed to be identified for

rural tourism. These destinations should be identified in regions that are underexplored for better resource utilization (Valeri 2016).

- Demographics of HP: Demographics of HP such as population, population density, literacy rate, and sex ratio will also influence the typology of tourism that can be developed. The divide in urban and rural regions will play its part.
- Economic incentives: Economic incentives act as a useful sales promotion tool. It can act as a potential means of diverting tourist flow via altering tourism products by making it more economical, like offering a bundle of products or offering lucrative discounts that can divert tourist flow to underexplored tourist areas. By providing attractive tour packages or giving lucrative discounts, tourist flow could be diverted to rural tourism (Kumar and Shekhar 2020b).
- Market segmentation for typology: How is the typology market segmented, targeted, and positioned will decide whether or not the product becomes successful. The segmentation should be measurable, substantial, actionable, and identifiable (Kumar an Shekhar 2020a).
- Value and culture of HP: The value and culture of the residents define their hospitality value. Also, it influences the typology of tourism that can be developed in the state.
- Digitalization in industry: Digitalization helps tackle the resource constraint, aligns with sustainability objectives, and adds the value to the product offering (Valeri et al. 2020).
- Current performance: Sub-par current performance of the industry necessitates the need to develop new products. The new products developed are at par with the industry's mission and vision and must align with the resources available.
- Competitive advantage of industry: The competitive advantage of the industry guides its future strategy. An industry can never overlook its competitive advantage if it wishes to survive in the long run.

After the factors are identified and coded, the next step is preparing the self-structural interpretation matrix (SSIM). For this, the interrelationship between the factors is identified and coded in the form of V-A-X-O. If a factor drives the second factor, the relationship identified is coded as V; if the relationship between the factors is such that a factor is dependent on the other factor, it is indicated as A; if there is mutual interdependency among the factors, then the relationship is indicated as X; and no interrelationship among the factors is treated as O. The following table indicates the relationship identified (Table 7.8).

In the subsequent stage, the SSIM is converted into a binary form of zeroes and ones. All the V and X relationships are converted into one, and the relationships of A and O are converted into zero. The resultant matrix obtained from this is known as the initial reachability matrix (Table 7.9).

From the initial reachability matrix, the next step is to incorporate transitive links into the matrix. The concept of transitivity here implies that if A factor leads to B factor and B factor leads to C factor, then A factor will lead C factor by the transitivity rule. After incorporating transitivity, the resultant matrix obtained is known as the final reachability matrix, and this further leads to the calculation of driving

Table 7.8 SSIM matrix of the factors

	01	02	03	04	05	06	07	08	09	10	11	12	13	14	15
01	V	V	V	O	A	A	A	A	O	V	A	A	O	O	A
02	A	V	V	A	A	A	A	A	A	A	A	A	A	A	A
03	A	A	V	A	A	A	A	A	A	A	A	A	O	A	A
04	A	O	V	V	A	A	A	A	O	O	O	A	O	A	A
05	V	V	V	V	V	A	A	A	A	V	O	A	V	V	V
06	V	V	V	V	V	V	A	O	A	V	V	A	V	V	V
07	V	V	V	V	V	V	V	V	A	V	V	A	V	V	V
08	V	V	V	V	V	O	A	V	A	V	V	A	V	V	V
09	O	V	V	O	V	V	V	V	V	O	O	A	V	V	V
10	A	V	V	O	A	A	A	A	O	V	A	A	A	A	A
11	V	V	V	O	O	A	A	A	O	V	V	A	O	A	A
12	V	V	V	V	V	V	V	V	V	V	V	V	V	V	V
13	O	V	O	O	A	A	A	A	A	V	O	A	V	A	V
14	O	V	V	V	A	A	A	A	A	V	V	A	V	V	V
15	V	V	V	V	A	A	A	A	A	V	V	A	A	A	V

Source: Author's elaboration

Table 7.9 Initial reachability matrix

	01	02	03	04	05	06	07	08	09	10	11	12	13	14	15
01	1	1	1	0	0	0	0	0	0	1	0	0	0	0	0
02	0	1	1	0	0	0	0	0	0	0	0	0	0	0	0
03	0	0	1	0	0	0	0	0	0	0	0	0	0	0	0
04	0	0	1	1	0	0	0	0	0	0	0	0	0	0	0
05	1	1	1	1	1	0	0	0	0	1	0	0	1	1	1
06	1	1	1	1	1	1	0	0	0	1	1	0	1	1	1
07	1	1	1	1	1	1	1	1	0	1	1	0	1	1	1
08	1	1	1	1	1	0	0	1	0	1	1	0	1	1	1
09	0	1	1	0	1	1	1	1	1	0	0	0	1	1	1
10	0	1	1	0	0	0	0	0	0	1	0	0	0	0	0
11	1	1	1	0	0	0	0	0	0	1	1	0	0	0	0
12	1	1	1	1	1	1	1	1	1	1	1	1	1	1	1
13	0	1	0	0	0	0	0	0	0	1	0	0	1	0	1
14	0	1	1	1	0	0	0	0	0	1	1	0	1	1	1
15	1	1	1	1	0	0	0	0	0	1	1	0	0	0	1

Source: Author's elaboration

power and the dependence level of each factor. The driving power of a factor indicates the total of the factor leading or driven by a particular factor, and dependence level indicates the total of factors on which a factor depends (Table 7.10).

The next step is to determine the partition level of each factor in the final ISM digraph. This step requires the use of reachability set, antecedent set, and intersection set for each factor. The reachability set of a factor includes those factors, including it, driven by the factor, and the antecedent set includes all those factors, including itself, on which the factor depends. Intersection set comprises factors which are

Table 7.10 Final reachability matrix

	01	02	03	04	05	06	07	08	09	10	11	12	13	14	15	Driving power
01	1	1	1	1	0	0	0	0	0	1	0	0	0	0	0	**5**
02	0	1	1	0	0	0	0	0	0	0	0	0	0	0	0	**2**
03	0	0	1	0	0	0	0	0	0	0	0	0	0	0	0	**1**
04	0	0	1	1	0	0	0	0	0	0	0	0	0	0	0	**2**
05	1	1	1	1	1	0	0	0	0	1	0	0	1	1	1	**9**
06	1	1	1	1	1	1	0	0	0	1	1	0	1	1	1	**11**
07	1	1	1	1	1	1	1	1	0	1	1	0	1	1	1	**13**
08	1	1	1	1	1	0	0	1	0	1	1	0	1	1	1	**11**
09	1*	1	1	1*	1	1	1	1	1	1*	1*	0	1	1	1	**14**
10	0	1	1	0	0	0	0	0	0	1	0	0	0	0	0	**3**
11	1	1	1	0	0	0	0	0	0	1	1	0	0	0	0	**5**
12	1	1	1	1	1	1	1	1	1	1	1	1	1	1	1	**15**
13	0	1	0	0	0	0	0	0	0	1	0	0	1	0	1	**4**
14	0	1	1	1	0	0	0	0	0	1	1	0	1	1	1	**8**
15	1	1	1	1	0	0	0	0	0	1	1	0	0	0	1	**7**
Dependence Level	**8**	**13**	**14**	**9**	**6**	**4**	**3**	**4**	**2**	**11**	**7**	**1**	**8**	**7**	**9**	

Source: Author's elaboration

Table 7.11 Partition level of the factors

	Dependence level	Driving power	Partition level
01	9	5	Level IV
02	13	2	Level II
03	14	1	Level I
04	10	2	Level II
05	6	9	Level IX
06	4	11	Level X
07	3	13	Level XI
08	4	11	Level X
09	2	14	Level XII
10	12	3	Level III
11	8	5	Level V
12	1	15	Level XIII
13	8	4	Level VII
14	7	8	Level VIII
15	9	7	Level VI

Source: Author's calculation

present in reachability and antecedent set. If the reachability set of an intersection set of a factor is the same, then that factor is eliminated, and its partition level is determined. This iteration is carried out until the partition level of each factor is determined (Table 7.11).

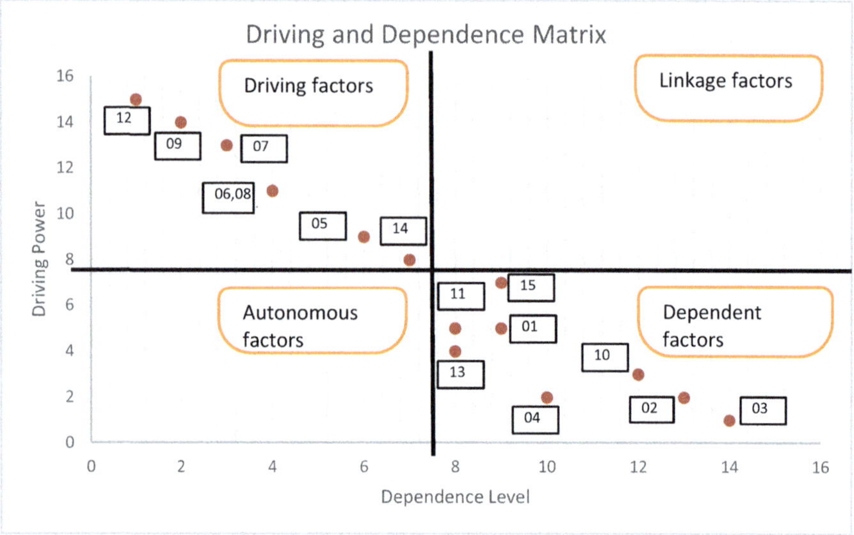

Source: Author's calculation

Fig. 7.5 Driving and dependence matrix. (Source: Author's calculation)

Conclusion from Driving and Dependence Matrix:

- Autonomous factors: These factors are characterized as having a low dependence level and low driving forces as well. These factors do not depend on other factors in the system and do not cause significant influences themselves. No autonomous factor is identified in our model (Fig. 7.5).
- Driving factors: These factors are characterized by high driving power and low dependence level. These are the change bringers in the framework. Any change in these factors causes an overall change in the system. Value and culture of HP, demographics of HP, resource availability, focus on sustainability, destination tourism potential, current performance of industry, and infrastructure development are the driving factors in our model.
- Linkage factors: Linkage factors are those factors which have high driving power and high dependence level as well. These factors are responsible for transferring the change in driving forces to the dependent forces by acting as a linkage between them. No linkage factor is identified in our model.
- Dependent factors: These factors are characterized by having a high dependence level and low driving power. These factors absorb the changes from all other factors and help in achieving the objectives. Digitalization, competitive advantage, market segmentation of typology, unique USP, economic incentives, marketing of typology, willingness of supply side to serve, and willingness of tourism to come are the model's dependent factors.

Once the partition levels are determined, the next step is preparing the digraph and establishing the links between these factors. The initial digraph (Fig. II) is prepared and represents all the links that exist between the factors. This shows the hierarchical relationship, which is established between the factors. Fig. III shows the final hierarchical relationship between the factor and is drawn to eliminate all the transitive links (Figs. 7.6 and 7.7).

Rural tourism would endure only if people are willing to consume it. There are plenty of studies highlighting the influence of rural tourism on the region's economy and social development, but studies are limited on whether tourists in India want to consume rural tourism product. The segment, if not substantial, would not lead to any economic good. This study also highlights a specific gap in the rural tourism research, which could provide researchers with issues to ponder. There are plenty of activities from which revenue could be generated in rural tourism. However, the problem is on setting up the pricing of rural tourism activities. The price should not be such high that it disinterests the tourists, and it should not be such low that no gains are generated from it. Researchers must undertake a study to frame the pricing mechanism for rural tourism. Although tourism is a service-based industry, setting up standards is an uphill task, but researchers must identify and set up specific basic standards that the destinations under the rural tourism umbrella must meet. The standards should be carefully devised in collaboration with the inputs from other stakeholders. The government should build up public forums to discuss the experiences, issues, and praises of tourists visiting rural areas. This would help them connect better with the tourists, understand and adapt to their wants, and provide a platform where people can recommend each other. Separate boards and committees should be constituted for the promotion and development activities of rural tourism. This would also ensure the local community's support, which would boost the development of the industry as a driving force.

Also, plans made in the tourism industry are rigid. If the plans are oriented toward a problem, they become obsolete and waste when the problem is solved, thereby requiring formulating new plans. Planning in tourism should be such that it is self-sustaining and can accommodate and adjust to problems and opportunities entering into and exiting from the system. This study attempted to provide different ways to stimulate the growth of the rural tourism industry. The study presents many feasible and possible channels to kick off the development in the rural tourism industry. A suitable channel could be selected to boost industrial development by coupling this analysis with high-medium-low and vital-desirable-burden analysis. There is cut-throat competition in the industries today. However, there is a requirement of coordination between allied and core industries of the tourism and hospitality sector. Small businesses can play a vital role in the development of rural tourism.

A core group or working committee could be framed, comprising of members from all entrepreneurs from all concerned sectors to work to better the industry. The rural tourism industry could not be developed in isolation in one or two states. As rurality is present throughout the country, planning agencies could utilize this advantage to develop integrated rural tourism-based circuits that connect rural tourism-based destinations and incorporate tourism activities such as wellness

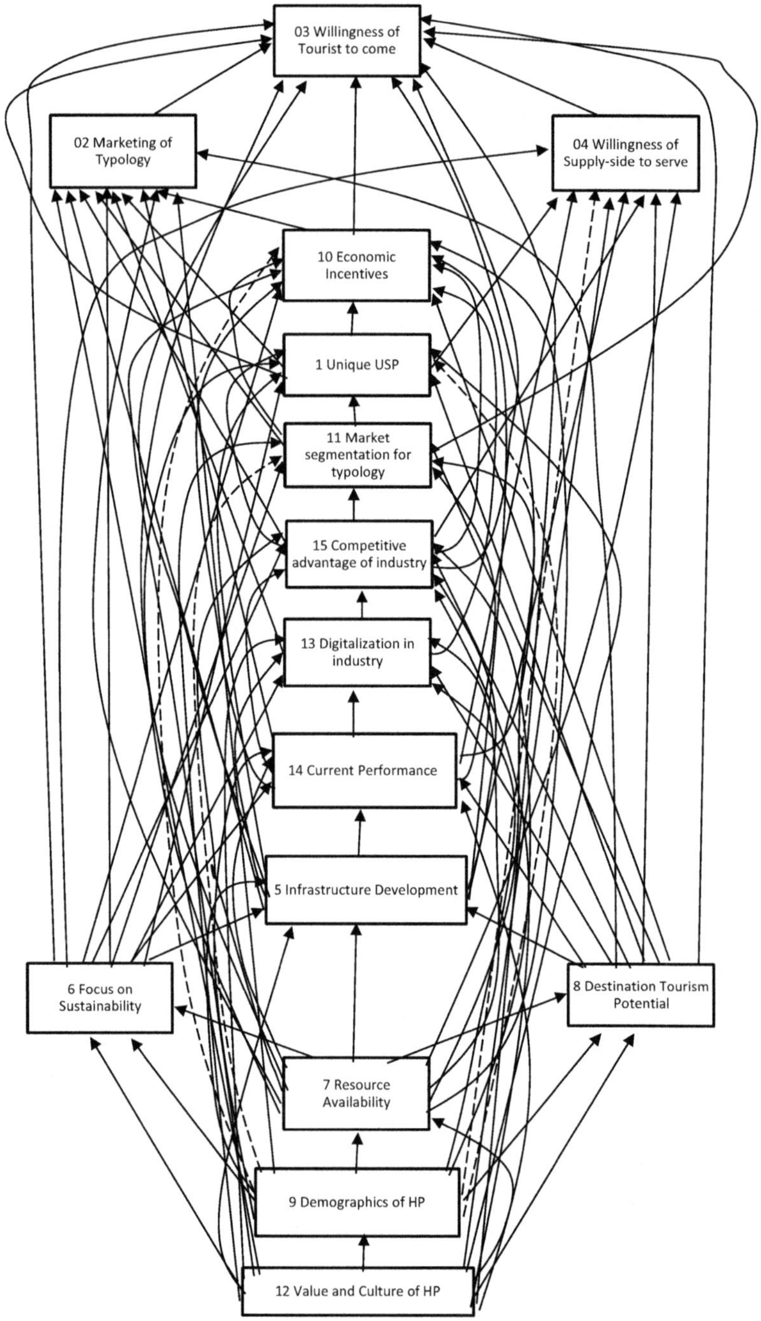

Fig. 7.6 Initial digraph showing hierarchy of factors. (Source: Author's representation)

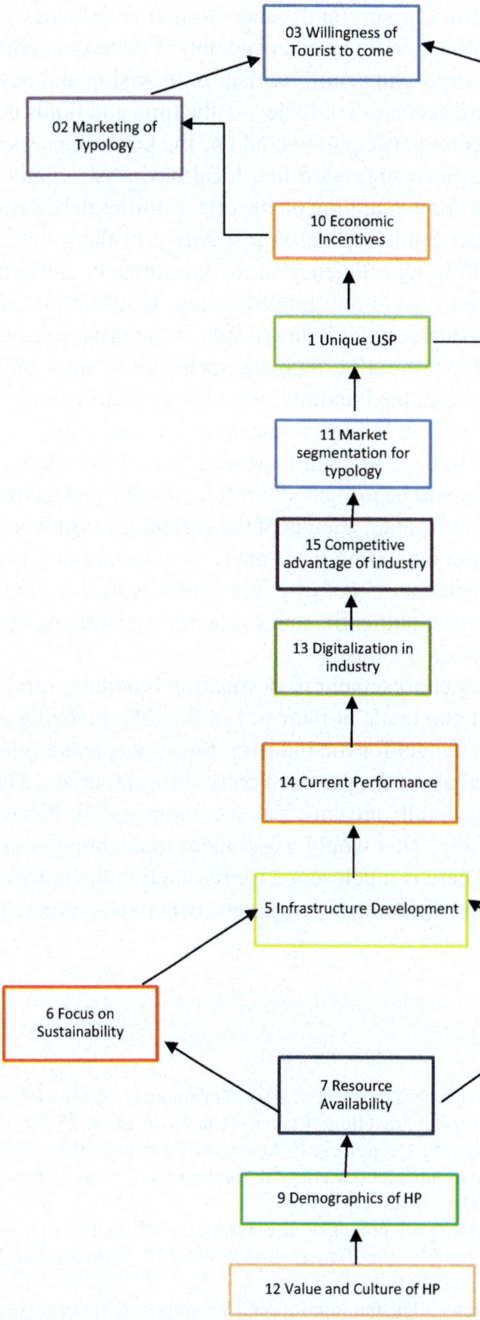

Fig. 7.7 Final digraph showing hierarchical relationship between the factors. (Source: Author's representation)

tourism and ecotourism, agro-tourism, and other similar typologies. Product requirement demands focus, which could be achieved only if there is a commitment from the organization. By incorporating rural tourism in its vision and developing missions to achieve it, planning agencies could devote the time and funds necessary for its development. Private sector participation could be the key to success for rural tourism development. Studies have suggested that local businesses blame lack of support from the private sector for not tapping on the opportunities developed. Since infrastructure development and funds allocation rest solely in the public sector's hands, the private sector could bring efficiency in its development and utilization. Strategic alliances between the private and public sectors would bring necessary cooperation between small businesses and industrialists in the private sector. Rural tourism could pave the way for promoting financial inclusion in rural India. With the system getting more organized, the handling of monetary transactions could be revolutionized. The use of digitalization in payment would further boost the process of financial inclusion in rural India. Although there is a lack of knowledge among the rural population, the issue could be mitigated through training and technical sessions. Social inclusion is one of the main agenda of the current government of India. Underexplored areas and people inhabiting these areas are unaware of the development in the world. These people can benefit by interacting with the communities from outside the region. A robust, authentic, and exclusive marketing campaign is needed to promote the rural tourism industry.

Moreover, through celebrity endorsement or destination branding, rural tourism marketing could be enhanced and made attractive. For this, the planning agencies need to increase the allocated budget for the industry. Since people are reluctant to change, they must be informed about the pros and cons of rural tourism. They must also be informed about what generally are anticipated of them and the behavior they need to improve their hospitality. This would assist them in adapting better to the needs and wants of tourists. There is much scope for research in the tourism industry. Since rural tourism is new to the country, the government must promote research in the area.

References

Chemli, Samiha, Michail Toanoglou, and Marco Valeri. 2022. The impact of Covid-19 media coverage on tourist's awareness for future travelling. *Current Issues in Tourism* 25 (2): 179–186.

Fernández-Hernández, Carlos, Carmelo J. Leon, Jorge E. Araña, and Flora Díaz-Pére. 2016. Market segmentation, activities and environmental behaviour in rural tourism. *Tourism Economics* 22 (5): 1033–1054. https://doi.org/10.5367/te.2015.0476.

Kachniewska, M.A. 2015. Tourism development as a determinant of quality of life in rural areas. *Worldwide Hospitality and Tourism Themes* 7 (5): 500–515. https://doi.org/10.1108/WHATT-06-2015-0028.

Kumar, S., and Shekhar. 2020a. Technology and innovation: Changing concept of rural tourism-a systematic review. *Open Geosciences* 12 (1): 737–752. https://doi.org/10.1515/geo-2020-0183.

———. 2020b. Digitalization: A strategic approach for development of tourism industry in India. *Paradigm* 24 (1): 93–108. https://doi.org/10.1177/0971890720914111.

Kumar, S., K. Attri, and Shekhar. 2018a. Niche tourism: Sustainability and marketing strategies. *Tourism Innovations* 8 (01): 1–9.

———. 2018b. Strategies for holistic development of tourism industry in India: A case study of Himachal Pradesh. *Journal of IMS Group* 15 (2): 12–22.

Kumar, S., N.G. Shekhar, and N. Guleria. 2019. Understanding dynamics of niche tourism consumption through interpretive structure modeling. *Saaransh: RKG Journal of Management* 11 (1): 40–48.

Lal, M., S. Kumar, and Shekhar. 2019. Understanding the dynamics of length of stay of tourists in India through interpretive structure modeling. *Review of Professional Management* 16 (2): 14. https://doi.org/10.20968/rpm/2018/vl6/i2/141019.

Osti, L., and L. Cicero. 2018. Tourists' perception of landscape attributes in rural tourism. *Worldwide Hospitality and Tourism Themes* 10 (2): 211–221. https://doi.org/10.1108/WHATT-12-2017-0087.

Valeri, M. 2016. Networking and cooperation practices in the Italian tourism business. *Heritage & Services Marketing* 2 (1): 30–35. https://doi.org/10.5281/zenodo.376333.

Valeri, Marco, Rosario Fondacaro, Cinzia De Angelis, and Andrea Barella. 2020. The use of cryptocurrencies for Hawala in the Islamic Finance. *European Journal of Islamic Finance.*

Himalayas, Climate Change, and Vulnerability to Society

8

8.1 Introduction

The word "Himalayas" has its roots in Sanskrit and essentially translates to "the abode of snow." The theory of continental drift recommends that formation of the Himalayas occurred about 40–50 million years ago, at the time when Eurasian plates were trenched with the Indian tectonic plates (Kious and Tilling 1996). The mountain range Himalayas is the tallest in the world and is outstandingly the home to ten of the fourteen most noteworthy pinnacles in the world (Godin et al. 1999). The prestigious Himalayan region is recognizably impacted by the climate change. The delicate landscapes of the Himalayan area are exceptionally vulnerable to common dangers, and there is continuous worry about present situation and likely climate change impacts which might incorporate strange floods, dry seasons and avalanches (Barnett et al. 2005; Cruz et al. 2007), dangers to food security, and the loss of biodiversity (Xu et al. 2009). The most generally detailed impact is the hastened melting of glaciers, which has significant future suggestions for downstream water assets. The Himalayan tropical/subtropical climatic conditions are portrayed from the foothills of the mountains to snow-covered peaks and perpetual ice at high elevations (Pant et al. 2018). The vegetation of the Himalayas fluctuates with environment, precipitation, elevation and soils. The measure of yearly precipitation increments along the southern front of the reach from the west to east. Besides, the Himalayas is depicted by various climate in sub-zones owing to different topographical changeability, which is firmly connected to topography (Bookhagen and Burbank 2006).The effects of environmental change are superimposed on an assortment of other numerous generally perceived as serious ecological and social burdens (Ives and Messerli 1989). Proceeding with environmental change is anticipated to prompt major changes influencing the environment of the Himalayan area. The district has an impressive hydropower potential and feeds various enduring streams which rely on the practical presence of the glaciers (GoI 2010; DST 2012). Different social groups are affected differently by climatic variability, with underprivileged

S. Kumar, *Sustainable Rural Tourism in Himalayan Foothills*,
https://doi.org/10.1007/978-3-031-40098-8_8

people being the most at risk. The majority of people in the Himalayas don't have the necessary connections to the outside world because they mostly reside in rural areas. As a result, food security suffers during unfavorable climatic conditions. Because of this, it is imperative to gain a thorough understanding of indigenous traditional food practices as well as cultural values, beliefs, and conventions in order to address the negative effects of environmental changes and emphasize the urgency of Himalayan societies.

8.2 Himalayan Topography and Its Linkages

The Indian Himalayan area is home to around 0.051 billion individuals, a significant number of whom practice agribusiness in diverse and fragile environments, including diverse forests. Because of its high organic and social cultural variety, the area has been recognized as one of the "biodiversity hotpots" by Gautam et al. (2013). There are variety of observed impacts on waterway streams, on natural hazards, and on the individuals and their jobs. The huge spatial degree of the Himalayas ranges across the eight Asian mainland nations and is the origin of significant frameworks for ten rivers (Sharma et al. 2019, HIMAP) providing potable water, water system, and power to about 1.3 billion individuals in Asia—almost 20% of the total populace (Rashul 2014). Before the end of the twenty-first century, the average cline of temperature over the entire of India is projected to ascend by roughly 4.4 °C in contrast with the 1976–2005 (Sanjay et al. 2020). The compilation is extraordinary, but the time span to analyze the accelerated impact to environmental change is too short to even think about coming to unequivocal end results. There is a reasonable knowledge gap on economic, social, and other sectorial impacts of environmental change in the Himalayas.

8.2.1 Himalayas and Rural Tourism

Rural tourism is widely described as the tourism-based activities occurring at a rural region. However, over time, authors raised a doubt on precisely what constitutes a rural area. According to the rich repertoire of literature, various characteristics that are commonly observed in the rural areas comprise agriculture as a prominent income source, low degree of populace density with proper subculture, and other social habits inside the area. As evident from the domain of studies conducted under the rural development, rural tourism has an immediate impact on rural improvement and numerous components like to set up micro, small, and medium enterprises, the entrepreneurial attitude, and infrastructure development thus playing a crucial role for the development in a rural region. Lone (2014) studied the strategic role of agriculture coupled with the diversification for the purpose of progress in the agriculture sector and thereof the rural development. Indolia and Prasoon (2015) mentioned various schemes of state governments, rural development, and other related Government of India (GoI) departments.

Garjola and Singh (2017) conducted an investigation into the challenges faced by villages in less developed rural areas, aiming to address both minor and major issues comprehensively. The study analyzed that the growing population level is the substantial obstacle in the path for development in rural regions. Government authorities should assist in controlling the increasing levels of the population, by informing rural people about family planning programs, practices, and policies. D. Vanitha and Dr. D. Vezhaventhan (2018) focused on the implementation of policies and the development within the rural regions. The study is further elaborated from the perspectives of the rural development; for the promotion of certain schemes, there is an urgent need to improve the facilities of Internet connectivity in the rural areas. Kapur (2019) studied the contribution and role of youth in the development of rural regions in India. The study highlighted that identifying the youth, their participation, obstruction of the youth participation, youth labor markets, vital role of youth demand in rural areas, and youth role assessment in the rural development are the prominent areas that would prove to be beneficial for rural development.

The literature suggests numerous potentials of the rural development in social and economic transformation of a region. Considerable measures need to be undertaken for the rural development to provide an opportunity for the business settings, utilization of technological innovations in operation, infrastructure development, etc. The abovementioned purposes are served by the rural tourism in the following ways:

- *Act as an alternative income source*: Majority of people in rural regions are dependent on earning their livelihood both on the farming and non-farming activities. Therefore, rural tourism can provide ample opportunities as an alternative source to livelihood. Various government-related schemes of knowledge awareness and skill acquisition for the promotion of livelihood aim to increase the earnings of people which could be used to impart knowledge among the stakeholders.
- *Bringing in employment opportunities*: The key characteristic of rural regions is the low assortment of employment opportunities. There exists little scope of possible employment available among these areas. Since the last few years, the level of unemployment has been troubling the government, and the rural tourism avenues could be exemplary options to overcome the employment gaps. Various government schemes like STRIVE (Skills Strengthening for Industrial Value Enhancement) scheme, promotion of entrepreneurship innovation, and agro-products scheme are some of the initiatives. The integration of such programs within the rural tourism sector holds the potential to play a pivotal role.
- *Sought in regional economic development*: The positive impacts of rural tourism development could be a springboard in regions which lack assets/resources to balance on economic and social fronts.
- *A means for social inclusion*: Rural tourism can play an important role in stimulating/creating awareness about the regional local traditions and customs. It can increasingly also be viewed as a mechanism to add locals in learning of the changes in lifestyle of large living population beyond their habitual zone.

- *To tackle social evils*: Rural tourism shall be able to provide a channel to overcome existing societal evils. During a respondent rural visit, an observation is focused on how rural tourism can be utilized as a means in receipt of providing relief to orphan children and support to elderly individuals residing in old-age homes seeking attention from their well-wishers.

The gradual shifting burden due to the exhausting destinations: Around India, several tourism destinations are significantly depleted and various natural assets at their dispersal are robustly depleted. Hence, the development in rural tourism sector can act as a center stage to redirect progression at destinations by tourists with comparatively lesser footfall and high resources, in a way that debilitating locations could get time to replenish. A significant literature spins around the conceptualization and significance of rural tourism. There is an inherent need to identity the factors that are exhibited for the successful rural tourism with the help of local communities, groups, and entrepreneurs. Adequate funding and local government support and good coordination, communication, and cooperation between entrepreneurs at various levels can be perceived as significant driving components for rural tourism. By inviting society at large, rather than imparting knowledge in classrooms, there could be a reciprocation of traditions, customs, and knowledge assisting for a preferable social inclusion. Most ideal ways the tourism industry can be utilized to advance social consideration all the more adequately and the understanding the impacts of rural tourism as well as its outcomes on to social inclusion, thereby adding to the rural development. The community-centric approach development can help in integrated rural tourism development. This perspective suggests that the involvement of youth holds substantial significance in propelling the advancement of rural tourism.

8.2.2 Himalayas and Global Climate Change

Climate change (CC) is a significant challenge confronting our planet earth. This is a comprehensive danger that will present critical economic, environmental, social, and political difficulties in the coming decades. The global climatic change has arisen as a worldwide ecological issue which has connected with the world consideration as it identifies with atmosphere globally. The long-term impacts are probably going to adversely influence the vulnerable population, whose contribution is least toward the significant reasons for climate related changes. Also concerning the early warnings of climate change, mountains are among the indicators (Singh et al. 2010). As glacial masses retreat, and snowlines move upward, waterway streams are probably going to change, and adjustment in water stream system might result in a plenty of social issues and influence generation of hydropower, impacting forests' imperil biodiversity, horticulture-based occupations, and well-being of the individuals.

Environmental change and a dangerous atmospheric deviations have become significant themes for established researchers and media lately. The convergences of ozone harming substances, for example, carbon dioxide levels in the climate, which ingest active long-wave radiations and are liable for a dangerous atmospheric deviation, have been expanding drastically. In spite of the fact that proof for worldwide environmental change has been grounded by mainstream researchers all throughout the planet, the impacts of these progressions on rural environments actually should be explored. The influence of these alterations in the environment is intricately shaped by regional-scale factors driven by local geographic characteristics and compounded by land-use patterns. For instance, the exceptional orographical highlights of the Indian subcontinent, for example, the giant mountain range of Himalayas, can affect environmental change. Since we are worried about the outcomes of an Earth-wide temperature boost and environmental change later on, it is essential to examine the impact of worldwide environment changes on the local scale. These models are integrated with global circulation models, as demonstrated in the work of Giorgi and Mearns (1991). A few dynamical and factual models came to be utilized in environmental change concentrates over the Himalayan region (Bajracharya et al. 2007; Marke and Hank 2009; Chaujar 2009; Shrestha 2009).

The changing environment and atmospheric warming generally affects water assets, agribusiness, and the general economy of the country. Varieties in the recurrence and force of outrageous climate occasions, like floods, dry spells, tornadoes and western unsettling influences, and the event of climate occasions, for example, cold and warmth waves over India permit established researchers to inspect whether boundaries related with these occasions can be regarded as proof for environmental changes in India. Hence, it is important to contemplate various parts of environmental change, in view of the accessible long haul information over various district.

8.2.3 Observed Changes in Response to Climate Change

A few observational examinations show huge precipitation and temperature changes in India over the last decades. A new report showed an expansion in the entire Indian data of the mean yearly temperature by 0.5 °C from 1901 to 2003 (Kothawale and Rupa Kumar 2005). This increase is predictable with the normal impacts from an Earth-wide temperature boost. The temperature pattern during the period of March–May observed concentration of pre-storm over the regions of western Himalayas as reported by Yadav and others (2004) utilizing perceptions and recreations with the help of tree rings. They tracked down the pre-rainstorm least temperature that would in general diminish during the late twentieth century. They likewise tracked down that the pace of diminishing of least temperature is multiple times with that of the pace of abatement of greatest temperature, showing that the base temperature is a bigger supporter of the cooling pattern in the pre-storm mean temperature. The temperature and precipitation patterns across northern and southern India during distinct phases of the monsoon are explored in previous studies (Dash and Hunt 2007; Dash et al. 2007). They discovered huge contrasts in patterns in least temperature and overcast cover among north and south India and lop-sidedness in expanding temperature patterns between various seasons. Likewise, they discovered a decrease in the late spring storm (June–September) precipitation in India and recommended that this decrease can be clarified by both environmental changes and the meso-scale mountain impacts on rainstorm stream. Long haul patterns in the most extreme, the least, and the mean temperatures in the twentieth century over the Himalayan northwestern region (Bhutiyani et al. 2007) propose a huge rise in temperature of northwestern Himalaya, as the arrival of winter warmth is happening at a quicker rate. The examination likewise shows that critical warming began in the last part of the 1960s, with the most elevated pace of increment somewhere in the year 1990 and 2009. Dimri and Ganju (2007) recreated precipitation and winter temperature over the western Himalayas. They utilized a provincial environment model and deciphered that temperature is thought little of and overestimated precipitation levels in the Himalayas. Notwithstanding, concentrates by Fowler and Archer (2006) showed some clashing outcomes. Since the 1960s, as per utilizing Upper Indus Basin (UIB) information, a diminishing pattern in mean yearly temperature was observed, which has gotten more articulated since the 1970s. Interestingly, they tracked down an expanding pattern during December–February (mean winter) temperature during the time period from 1960 to 2000. The enduring patterns of precipitation and temperature over the western Himalayas were analyzed, and the series of events of warm and cold throughout winters (December–February) for the 1975–2006 period were determined (Dimri and Kumar 2008). They discovered a pattern of expanding temperature and diminishing precipitation in particular areas. Observational examinations (Kripalani et al. 2003) additionally uncovered that the snow cover revealed a decline in spring across the western Himalayas and a quicker reduction in the snow from winter to spring since 1993, which might be because of a dangerous atmospheric deviation. To satisfy India's vision of economic advancement with regard to environmental changes, a Climate Change Public Action Plan was initiated in June 2008. The plan incorporates eight explicit goals, including of "National Mission on Strategic Knowledge for Climate Change" and "National Mission for Sustaining the Himalayan Eco-systems."

8.2.4 Resource Sustainability in Tourism

The wonder of the travel industry is the substantial aftereffect of travelers utilizing public assets like mountains, grasslands, forests, lakes, and mountains and their related leisure activities like trekking, setting up camps, water sports, making the travel industry supply a complex mix of natural assets, sporting destinations, and the travel industry's dependent organizations. The travel industry development offers a conceivably stable advancement choice in tourist destinations for the perspective on enhanced demand for leisure, other community-based advantages, and impression of the travel industry as a clean industry. With the travel industry items showcased in different structures (going from rural to urban tourism, adventure tourism cultural destinations to ecotourism and medicinal and health tourism), the travel industry utilization has changed broadly lately, making more fragmented and concentrated markets.

A critical inquiry that emerges during development of burgeoning rural tourism industry is that on the one hand, it focuses on over-tourism at specific spots, and on the other hand, on reporting sustainability. In this way, a lot of research has led to the best way to make country's rural tourism sustainable. The buzzword sustainable tourism has been characterized as a way to deal tourism activities with the respect to necessities among other stakeholders just as the requirements for future tourists, without compromising current tourist's satisfaction. Social inclusion is the way toward improving the terms on which individuals and gatherings participate in improvising the society and providing opportunities and dignity for individuals disadvantaged on the basis of their identity. Such endeavors incorporate approaches and activities that elevate equivalent admittance to public administrations just as empowering resident's investment in the dynamic cycles that influence their lives. Sustainable rural tourism, accordingly, can be perceived as a rural tourism that assumes no overburdening at destination as far as resource utilization is concerned and no loss of satisfaction among rural tourists. The well-known sustainability practiced continued in tourism is limiting the tourist aggregated at a tourist destination. The current approach is anyway reprimanded for being anti-stakeholders.

Numerous travel industry analysts have considered ecological assets' travel industry potential and how they trigger the travel industry's interest and carry worth to an objective. Environmental issues are presently more significant than any time before and are currently during COVID-19 pandemic as an indispensable part of tourism development. Therefore, to incorporate sustainability in tourism, industries need to adopt other strategies and practices.

8.2.5 Himalayas and Sustainable Rural Tourism Strategies

Mountains are distinctive ecosystems, distinguished by the unique types of diversity and variability they exhibit (Messerli and Ives 1997). Mountainous regions are drawing in extraordinary consideration among foreign and Indian sightseers. Key aspects in advancing ecotourism include environmental conservation, education, and disseminating strategies for achieving sustainable travel objectives. These efforts raise awareness about the value of natural resources and contribute to sustainable tourism goals. Steep geological, climatic, and natural inclinations joined with sharp seasonal differences favor the setting off of outrageous climatic and geomorphic occasions, which may thus influence biological and human conditions (Price 1999; Ives and Messerli 1989; Viviroli et al. 2007; Körner 2013). While 12% of the global populace reside in mountains, the incredible dominant part live in economically and physically marginalized developing nations, so that an environmental shift shall impact on the rise of poverty and issues of food security (Kohler et al. 2014). Mountain populaces have a variety of lifestyle: traditional activities, like farming, forestry, and animal husbandry, exist together with mining explorations and creation of hydropower and other tourism-related events (Beniston et al. 1996). Mountains, likewise, are fragile and they can be damaged by various anthropogenic activities like deforestation, development projects on peripheral soils, overgrazing by livestock, and development of urbanization, all of which may cause first speedy deterioration of biodiversity and aquatic resources and second an extension in characteristic perils, henceforth putting populaces at greater risk of danger.

Given that Himalayan objections have their own restricted regular and legacy assets and foundation to complete the travel industry advancement, the executives of destination carrying capacity becomes indispensable. Control components should be nearby setting explicitly yet inescapably to make the "carrying capacity concept" stick. Each state can foster its standard arrangement of norms to characterize carrying capacity in light of rules for maintaining the travel industry once on a transient premise the idea is tried. Physical framework, for example, lodgings, eateries, and street and trail networks in backwood regions, should be based on naturally well-disposed nature. Clear drafting of scene (land use plans) will permit tweaked framework advancement which will guarantee that out of line and negative compromises are stayed away from. Moreover, such a drafting will help observe and guide site explicit mediations if and when required. Sustainable travel industry models likewise should be produced for provincial travel industry, homestays, and other travel industry administrations.

8.3 Conclusion, Challenges, and Future Directions

Climatic endurance and adaptation rely upon various biological systems and associated local patterns. Environmental changes can have detrimental effects on the living entities and their nearby climate, decreasing survival and propagation and causing impacts on populaces or species' dispersions across geographic locales. Changes in environment may benefit a few animal varieties but may harm people. Altogether, it will modify natural processes and functioning of biological systems. Changes in the capacities of various biological system can thus increase or decrease the pace of human-driven natural change. Notwithstanding impacts of environment factors like temperature levels and rate of precipitation, plants may react directly to the rising convergence of CO_2, while faunal species in aquatic ecosystems acclimatize to water changes due to dissolution of greenhouse gases. The earth is currently experiencing sufficient ecological change to impact characteristic structures; all around announced changes in plant and animal peoples are related to late environmental changes. Foreseeing future biological implications of climate change stays a considerable challenge:

- Environment affects life-forms: plants, animals, creatures, populaces, networks, and the entire biological systems. Changes in environment will have sweeping impacts on all parts of science.
- Climatic changes in the recent years have effectively created quantifiable changes in organic frameworks around the world. Species can respond to natural change by moving to regions where the environment is good for their health, by advancing and adjusting to new ecological conditions or by progressing and acclimating to new environmental conditions.

- Investigating ongoing patterns ensures that changes will happen in natural frameworks esteemed by humans. A large number of those progressions will contrarily affect human prosperity, yet there will be changes that will likewise benefit individuals. Researchers attempt to foresee how natural systems will change by dissecting past on account of ecological changes, by directing analyses and by developing models.
- Vulnerability in expectations of biological changes comes both from vulnerability about the pace of future environmental change and from vulnerability about the immediate and roundabout impacts of environmental change on natural frameworks. The incorporation and execution of a "Green cess" should likewise act as the reason for out-scaling and up-scaling of tourism-based activities.

Socioeconomic turn of events is progressively tested by the mass travel industry situation, which has likewise a cross-line interface and countenances monstrous environmental change-related difficulties, as well as the regular issues of enormous monetary shortages, holes in natural consistency, and very negligible ventures to advance feasible travel industry. These issues can deteriorate if the visualized traveler heap of 240 million by 2025 turns into a reality. Indeed, even states with best practices in the travel industry (e.g., Sikkim) are similarly undermined by mass travel industry and its fallouts, aside from losing backwood cover to advancement. Socioeconomic transformation is steadily confronted by the ramifications of mass tourism, which extends beyond domestic borders and grapples with substantial climate change-related trials. Furthermore, this scenario is compounded by persistent issues like significant economic deficits, disparities in ecological integrity, and limited efforts to foster sustainable tourism. These issues can deteriorate if the visualized traveler heap of 240 million by 2025 turns into a reality. Indeed, even states with best practices in the travel industry (as are Sikkim) are similarly undermined by mass the travel industry and its fall-outs, aside from losing backwoods cover to advancement. To accomplish the goal of community participation and social incorporation viably, development of rural tourism might play an instrumental role. But the principal concern is formulated policies and strategies for the conducive rural tourism development. There can be certain benefits from socioeconomic contexts in relation to the rural tourism. How effectively the probable benefits shall be flushed in shall determine how rigorously one will be able to achieve the social inclusion. Alongside improving comprehensiveness and connectedness in the society, one may accomplish development in rural pockets, which perhaps likewise can positively impact the development in the tourism sector. But regarding the treasures of rural tourism, the key issue is to achieve sustainability. Sustainability will corroborate on how one can sufficiently accomplish all the above-tended proclamations. This will empower the tourism industry to proceed over the long haul and will likewise empower the framework to retain issues which are not existential at present but rather can be complex to deal in the later near future.

Further, the state planning commissions should guarantee and manage intermingling inside the travel industry divisions, between various areas to evaluate and direct plenty of economical travel industry viewpoints and related data, viz., limit building, advertising, and advancement including item improvement, principles, affirmation and rules, executive information and examination, and coordinated objective and framework arrangement and execution. A follow-up framework advancing the tourism industry and its packages (homestays, rustic travel industry, executive legacy) should be mixed inside the educational plan furthermore and conveyance plan of preparation phases/instruction establishments (e.g., professional focuses, colleges, schools) serving a large group of entertainers and public/confidential areas. There is a need to release this data through a proactive media crusade and existing travel-related sites. Comprehensive travel industry satellite records should be made accessible as "one window online data" (e.g., data on the number of lodgings/rooms, traveler locales, stopping places, traffic reconnaissance, squander quantum, number of local area experts, consumptions/influences made on advances supporting the travel industry).

References

Bajracharya, S.R., P.K. Mool, and B. Shrestha. 2007. *Impact of climate change on Himalayan glaciers and glacial lakes: case studies on GLOF and associated hazards in Nepal and Bhutan.* Kathmandu, International Centre for Integrated Mountain Development and United Nations Environment Programme Regional Office, Asia and the Pacific. (ICIMOD Publication 169.)

Barnett, Tim P., Jennifer C. Adam, and Dennis P. Lettenmaier. 2005. Potential impacts of a warming climate on water availability in snow-dominated regions. *Nature* 438: 303–309.

Beniston, M., D.G. Fox, S. Adhikary, R. Andressen, A. Guisan, J. Holten, J. Innes, J. Maitima, M. Price, and L. Tessier. 1996. The impacts of climate change on mountain regions. In: *Climate change 1995: impacts, adaptions, and mitigation of climate change: scientific-technical analyses.* Contribution of Working Group II to the Second Assessment Report for the Intergovernmental Panel on Climate Change [Watson, R.T., M.C. Zinyowera, and R.H. Moss (eds.)]. Cambridge University Press, Cambridge, and New York, pp. 191–213.

Bhutiyani, M.R., V.S. Kale, and N.J. Pawar. 2007. Long-term trends in maximum, minimum and mean annual air temperatures across the northwestern Himalaya during the twentieth century. *Climatic Change* 85 (1–2): 159–177.

Bookhagen, B., and D.W. Burbank. 2006. Topography, relief, and TRMM-derived rainfall variations along the Himalaya. *Geophysical Research Letters* 33 (8).

Chaujar, R.K. 2009. Climate change and its impact on the Himalayan glaciers – a case study on the Chorabari glacier, Garhwal, Himalaya, India. *Current Science* 96 (5): 703–707.

Cruz, R.V., H. Harasawa, M. Lal, S. Wu, Y. Anokhin, B. Punsalmaa, Y. Honda, M. Jafari, C. Li, and N. Huu Ninh. 2007. Asia, climate change 2007: Impacts, adaptation and vulnerability. In *Contribution of working group II to the fourth assessment report of the intergovernmental panel on climate change*, ed. M.L. Parry, Canziani OF, J.P. Palutikof, P.J. Van Der Linden, and C.E. Hanson, 469–506. Cambridge: Cambridge University Press.

Dash, S.K., and J.C.R. Hunt. 2007. Variability of climate change in India. *Current Science* 93 (6): 782–788.

Dash, S.K., R.K. Jenamani, S.R. Kalsi, and S.K. Panda. 2007. Some evidence of climate change in twentieth-century India. *Climatic Change* 85 (3–4): 299–321.

Dimri, A.P., and A. Ganju. 2007. Wintertime seasonal scale simulation over western Himalaya using RegCM3. *Pure and Applied Geophysics* 164 (8–9): 1733–1746.

Dimri, A.P., and A. Kumar. 2008. Climatic variability of weather parameters over the western Himalayas: A case study. In *Proceedings of the National Snow Science Workshop, 11–12 January 2008, Chandigarh, India*, Satyawali, P.K. and Ganju, A., eds. Chandigarh, Snow and Avalanche Study Establishment, pp. 167–173.

DST. 2012. *Dynamics of Glaciers in the Indian Himalaya: science plan*. Himalayan Glaciology Technical Report No. 2. Science and Engineering Research Board (SERB), Department of Science and Technology, New Delhi, p. 125.

Fowler, H.J., and D.R. Archer. 2006. Conflicting signals of climatic change in the upper Indus Basin. *Journal of Climate* 19 (17): 4276–4293.

Garjola, R., and K. Singh. 2017. Analytical study on the technological problems of rural India and their remedies. *International Journal on Emerging Technologies*: 623–624.

Gautam, Mahesh R., Govinda R. Timilsina, and Kumud Acharya 2013. *Climate change in the Himalayas current state of knowledge*. The World Bank Development Research Group Environment and Energy Team, Policy research working paper WP-6516, p. 47.

Giorgi, F., and L.O. Mearns. 1991. Approaches to the simulation of regional climate change: A review. *Reviews of Geophysics* 29 (2): 191–216.

Godin, L., et al. 1999. High strain zone in the hanging wall of the Annapurna detachment, Central Nepal Himalaya. In *Himalaya and Tibet: Mountain roots to mountain tops*, ed. A. MacFarlane, R.B. Sorkhabi, and J. Quade, vol. 328, 201. GSA.

GOI. 2010. *National Mission for sustaining the Himalayan eco-system under National Action Plan on climate change*. Draft Report. Department of Science and Technology, Government of India, p. 51.

Indolia, D.U., and K. Prasoon. 2015. An overview of policies & schemes of Govt. of India to promote rural sector & tourism. *International Journal of Applied Research*: 775–778.

Ives, J.D., and B. Messerli. 1989. *The Himalayan dilemma: Reconciling development and conservation*. London: Wiley.

Kapur, R. 2019. Role of youth in rural development. *ACTA Scientific Agriculture*: 298–301.

Kious, W.J., and R.I. Tilling. 1996. This dynamic earth: the story of plate tectonics. *USGS report*. https://doi.org/10.3133/7000097.

Kohler, T., A. Wehrli, andnM. Jurek (eds.), 2014. *Mountains and climate change: A global* concern. Sustainable mountain development series. Bernlin: Centre for Development and Environment (CDE), Swiss Agency for Development and Cooperation (SDC) and GeographicaBernensia. 136.

Körner, C. 2013. Alpine ecosystems. In *Encyclopedia of biodiversity*, ed. S.A. Levin, vol. 1, 2nd ed., 148–157. Amsterdam: Academic Press. https://doi.org/10.1016/B978-0-12-809633-8.02180-4.

Kothawale, D.R., and K. Rupa Kumar. 2005. On the recent changes in surface temperature trends over India. *Geophysical Research Letters* 32 (18): L18714. https://doi.org/10.1029/2005GL023528.

Kripalani, R.H., A. Kulkarni, and S.S. Sabade. 2003. Western Himalayan snow cover and Indian monsoon rainfall: A reexamination with INSAT and NCEP/NCAR data. *Theoretical and Applied Climatology* 74 (1–2): 1–18.

Lone, R.A. 2014. Agriculture and rural development in India; the linkages. *IMED*: 65–74.

Marke, T., and T. Hank. 2009. Using a regional climate model for the simulation of hydrologic processes in the high Himalayan Wangchu watershed. In *EGU General Assembly Conference Abstracts*, p. 11.

Messerli, B., and J. Ives, eds. 1997. *Mountains of the word. A global priority*, 496. New York: Parthenon Publishing.

Pant, Govind Ballabh, P. Pradeep Kumar, Jayashree V. Revadekar, and Narendra Singh. 2018. *Climate change in the Himalayas*. Springer. https://doi.org/10.1007/978-3-319-61654-4.

Price, M., ed. 1999. *Global change in the mountains*, 217. London: Parthenon Publisher.

Rashul, G. 2014. Food, water, and energy security in South Asia: A nexus perspective from the Hindu Kush Himalayan region. *Environ Science Policy* 39: 35. https://doi.org/10.1016/j.envsci.2014.01.010.

Sanjay, J., et al. 2020. Temperature changes in India. In *Assessment of climate change over the Indian region*, ed. R. Krishnan, J. Sanjay, C. Gnanaseelan, M. Mujumdar, A. Kulkarni, and S. Chakraborty. Singapore: Springer. https://doi.org/10.1007/978-981-15-4327-2_2.

Sharma, E., et al. 2019. Introduction to the Hindu Kush Himalaya assessment. In *The Hindu Kush Himalaya assessment—Mountains, climate change, sustainability and people*, ed. P. Wester, A. Mishra, A. Mukherji, and A.B. Shrestha. Berlin: Springer-Nature.

Shrestha, A.B. 2009. *Climate change in the Himalayas.* Kathmandu, International Centre for Integrated Mountain Development and United Nations Environment Programme Regional Office, Asia and the Pacific. (ICIMOD Publication 3/09.)

Singh, S.P., V. Singh, and M. Skutsch. 2010. Rapid warming in the Himalayas: Ecosystem responses and development options. *Climate & Development* 2: 1–13.

Vanitha, D., and D. Vezhaventhan. 2018. A study on rural development in Tamilnadu. *International Journal of Pure and Applied Mathematics*: 71–86.

Viviroli, D., H.H. Dürr, B. Messerli, M. Meybeck, and R. Weingartner. 2007. Mountains of the world, water towers for humanity: Typology, mapping, and global significance. *Water Resources Research* 43 (7): W07447.

Xu, Jianchu, R. Edward Grumbine, Arun Shrestha, Mats Eriksson, Xuefei Yang, Y.U.N. Wang, and Andreas Wilkes. 2009. The melting Himalayas: Cascading effects of climate change on water, biodiversity, and livelihoods. *ConservBiol* 23: 520–530.

Yadav, R.R., W.-K. Park, J. Singh, and B. Dubey. 2004. Do the western Himalayas defy global warming? *Geophysical Research Letters* 31 (17): L17201. https://doi.org/10.1029/2004GL020201.

Conclusion, Suggestions, and Study Limitations

9

9.1 Conclusion

This study was carried out to determine the industry's perspective concerning rural tourism development in Himachal Pradesh. Further, since there was a need to strategically develop the sector, this study applied flexible strategic management techniques to determine a clear path for industry development. Before arriving at the rural tourism development, it was also equally important to assess the tourism industry's present situation in Himachal Pradesh, India, and globally. The studies moved chronologically to develop an understanding of the key concepts in the study, understanding the work already done in the themes and sub-themes, finding research gaps and questions, creating a framework for the study, highlighting the tools and techniques to be used in the study, and analyzing the collected data to derive meaningful conclusions. Taking as a beginning stage the significance of the travel industry in the economy, one more helpful approach to recognize various advancements required in travel industry refers to the scale or effect of the travel industry. Likewise, one can see sensitization of the changes brought about in numerous country regions by the squeezing need to move on to another stage and to consider cautiously the potential and possibilities for future sustainable development. In light of the preparedness, several conclusions can be derived from the present study.

First, the study shows that the economic importance of the tourism industry is significant. The travel and tourism industry's financial contribution to global GDP and employment levels are significantly high and are rising over the years. There exist several challenges for the tourism industry at the international level, but these can be mitigated strategically.

Second, environmental scanning of Himachal Pradesh also raises the presence of several socioeconomic challenges in the state. The current status of the tourism industry is also raising an alarming situation. The typologies of tourism experienced are getting obsolete, which is visible in decreasing the tourists' footfall in the state. Popular destinations are being exhausted and soon will not be able to serve

S. Kumar, *Sustainable Rural Tourism in Himalayan Foothills*,
https://doi.org/10.1007/978-3-031-40098-8_9

additional tourists. The average length of tourist stay is also not encouraging and can be increased further. It is a severe concern for the state which is much dependent on tourism for its livelihood.

Third, there exist several unanswered questions that need to be answered for the significant development of the rural tourism industry. The role of technology, the extent of social and financial inclusion, and tourism in tackling such socioeconomic challenges are yet to be addressed by academia. Fourth, the status and trends in rural tourism in Himachal Pradesh are not encouraging. There is a lack of marketing for the product, and people are not aware of what it is and where to experience it. Further, there is a challenge to make tourism sustainable in the state without compromising the other stakeholders' economic gains. A solution must be derived that is mutually beneficial for the industry and is sustainable.

Fifth, the infrastructure development, growing environmental conscience, support of local government and community, availability of funds from the government, and participation of the private sector are the primary factors channelizing rural tourism development. Locals' attitudes to adapt, tourist travel motives, marketing of the destination, destination characteristics, and recommendation by others depend upon the primary factors and contribute to the rural tourism development. Concentrated efforts in making destination accessible, balancing between growth and sustainability, and support of local government are required for the industry's significant expansion.

Sixth, agro-based tourism products are in a much better position to be developed in rural areas. The rural areas have sufficient agro-based industries, and the farmers are themselves equipped with resources to serve agro-tourists. Along these lines, of the different typologies discussed in the study, agro-tourism takes the lead. Other niches can also be developed that have relatively high developmental potential and less market share at present. Seventh, the demographics of rural HP permits for the development of rural tourism. The educational level of the people also hints at the easy adoption of technology by the people. Female-friendly orientation also mitigates the risk posed to the rural tourism industry. Eight, the study concludes that there is a favorable prospect for developing the rural tourism industry in Himachal Pradesh. These prospects are visible to the policymakers, but stakeholders together believe that the development of rural tourism will be a positive sign for the industry and the state.

Ninth, the study concludes that climatic endurance and adaptation rely upon various biological systems and associated local patterns. Environmental change can disturb the match among living entities and their nearby climate, decreasing survival and propagation and causing resulting impacts on populaces' or species' dispersions across geographic locales. The earth is now encountering adequate environmental change to influence natural frameworks; all around reported changes in plant and creature populaces are identified with late environmental change.

9.2 Suggestions

In light of the discussion carried out in the study and research implication, several suggestions are made to build up a sustainable rural tourism industry. These suggestions could also be generalized to other states if the method and study are conducted out in other states. In a way, these suggestions would work for the rural tourism industry throughout the country. The recommendations are divided into sub-themes that were identified during the study.

9.2.1 Holistic Tourism Development Suggestion

The principal need to recognize is to ensure the holistic development of the tourism industry. Since there is an imbalance in the industry development, the desired results could not be achieved, even with the plans' strategic formulation and execution. To guarantee that there is the holistic development of the tourism industry, the authors make the following suggestions:

- Imparting tourism education: The travel industry education among young age ought to be mandatory. New courses, for instance, MBA in Tourism, ought to be advanced, and an appropriate educational plan ought to be confined to guarantee that such understudies get legitimate positions in the business. Moreover, research in the travel industry's territory ought to be empowered so new items could be created, remembering the requests of the vacationers. Exploration ought to likewise zero in on improving consumer loyalty and advancing maintainability in the business.
- Determining the carrying capacity: As examined, breaking of carrying capacity limit of a particular destination is the essential worry of policymakers. Along these lines, on the off chance that destination must be suspended, at that point, the carrying capacity limit of both ceased destination and new destination must be resolved. The carrying capacity limit ought not to be fixed uniquely in terms of the physical carrying limit. Social and economic carrying capacity also plays a pivotal role in determining demand at a destination.
- Development of air connectivity: What was seen is that destinations have air network pull in travelers unquestionably more in numbers than those having no such arrangement. Thus, as an answer, the government can anticipate building airstrips, so the air network is built up, particularly for areas such as Mandi, Chamba, and Kinnaur. Likewise, the overhauling of the existing air terminal is vital. Creating airstrips would also prompt the advancement of new travel industry roads in the state.
- Increasing tourists' stay in the state: Simple inflow of travelers is not anything to boast about. Vacationer appearance is valuable just when it is changed over into profit. For it, travelers are needed to remain in for a more drawn-out length. Various investigations have recognized components that sway the term of vaca-

tioner stays. What ought to be conceivable is understanding the elements of length of stay and planning adaptable systems for the equivalent.
- Circuit development in other districts: All the more overwhelmingly, center around creating the travel industry circuits has been in Kullu, Shimla, and Kangra. The organizations ought to distinguish the spots in a different locale and create theme-based tourism circuits along these lines. Examining travel destinations in Mandi reveals abundant opportunities for diverse visits, spanning religious, sightseeing, adventure, and wellness tourism. Along these lines, the offices ought to create circuits for underexplored places and market them successfully.

9.2.2 Improving Technological Infusion in the State

There is a need to improve the state's infrastructure to ensure a holistic development of the tourism industry. Technology will likewise assume a significant role in the development of the sector. However, for that, there is a need to boost up the technological infusion in the state. There are several ways to increase the technology infusion:

- Educating young stakeholders in the tourism industry: The young can go about as harbinger of progress. Likewise, if they are capable, youngsters can help in the spread of specialized instruction among kids and older folks. By utilizing innovation, they can increase the utilization of innovation by everybody engaged with the business. Along these lines, arranging organizations must guarantee that the young are taught with the utilization of innovation. For that, it needs to set up a foundation that bestows the specialized preparation and information.
- Protect and incentivize technology use: The arranging organizations must innovate the utilization the business by giving appropriate impetuses to the partners. These motivations can be as protected innovation rights, too, or as some financial limits and appropriations on innovation utilization in tasks. Laws identified with IPRs ought to be harsh, and unforgiving punishment must be enforced on the individuals who disregard these laws.
- Give financing needs to innovation execution: The absence of finance is likewise a significant issue that goes about as a boundary for technovation in the travel industry. In this manner, the administration or arranging organizations should set up assets to ensure that the absence of finance does not hinder the decent specialized advancement method. This initiative should assist any motivated individual in implementing innovative approaches within their tasks.
- Establish centers for modified technovation: Customization is fundamental to pick up an advantage over rivals. If businesses center a lot around creating innovation, they will lose their attention on their core work. The fundamental response to this issue is that a center must be set up for exclusively building up the tweaked specialized applications, so business and development go hand in hand.

- Safeguard against foreign firms: Advancement in the travel industry would consider foreign firms that approach high innovation and continually overhaul themselves because of high subsidy accessibility. Along these lines, the legislature must ensure the homegrown firms by forcing practical limitations on FDI and FII in the travel industry.
- Use technology to necessitate the use of technology: On the off chance that panning offices themselves start innovating and create stage for lawful consistence carefully, at that point, it would require the other stakeholders to innovate as well. It would be a demonstration of suasion without pressurizing the partners legitimately.

9.2.3 Strategic Development of Niche Tourism

The use of technology and the holistic development of the state will be incomplete without niche tourism development. The technological infusion will give rise to new products that will help develop the tourism industry in underexplored regions. Thus, to create and promote niche tourism, the policymakers should follow a few steps:

- Develop new tourism products: The travel industry is one of the industries where the rule of "Supply creates its own demand" can be applied. Along these lines, strategy creators must attempt to enhance their item's contributions and check whether a specific typology can be created as an appropriate travel industry item or not.
- Monitor trends and status of niche tourism products: A great deal of data hole exists relating to patterns and status of specialty travel industry items in the nation. Due to the absence of data, it is hard to outline arrangements and even have an appropriate examination. Along these lines, the job does not end just by building up the item, and the advancement ought to be checked, and obstructions must be assessed.
- Maintain affordability of tourism product: As the budget of tourists was one of the variables affecting the travel industry's specialty, special travel industry items ought to be valued cautiously. If costs are excessively high, it will demoralize the traveler from the travel industry. Too low costs would make it hard for partners to support over the long haul acquiring misfortunes. In this way, a parity must be kept up through a viable evaluation strategy.
- Identify tourists having a longer duration of stay: Specialty tourism could never turn into a full option to mass tourism. Alternatively, a vacationer would modify its stay at a popular location and spend a segment of its length on specialty travel industry. In this way, sightseers with longer-term stays are bound to embrace specialty in the travel industry. The center must be around such sightseers along these lines, and modified travel bundles could be offered to such vacationers.

9.2.4 Improving the Length of Tourist Stay Suggestions

Niche tourism development will, in a way, impact the length of tourist stay. As the length of stay of tourists can be increased in the state, policymakers can use several suggestions for the same:

- Development of theme-based tourism circuit: To have vacationers invest more time and money in the state, it is inescapable that a theme-based tourist circuit ought to be created. These circuits ought to be with the end goal that they associate everywhere in the state. It would likewise guarantee the comprehensive advancement of the travel industry in the state by assuring the spread of advantages from industry spans to the prohibited area.
- Development of new products: Travelers get exhausted with the same contributions. They need to encounter new and inventive travel industry items. Its repetitiveness decreases their length of stay. With an assortment of contributions to browse, they are glad to visit again, with their length of stay increased compared to their last visit. The offices should zero in on creating items which rely upon the attributes of the goal.
- Using a compelling marketing mix: Advertising is the most vital perspective where India's travel industry lingers behind. The advertising blend utilized ought to be with the end goal that it draws consideration of vacationer, build up a sentiment of enthusiasm for the location, build up a desire inside them to visit, and convince them to attempt the action of going on the outing.
- Focus on tourists with a longer length of stay: The business ought to go for a total picture redesigning. It would require a move in its vision and mission. Rather than zeroing in on travelers who have a place visit for a reason like MICE, the travel industry, where the length of stay is less, should change its concentration to health travel industry where the length of stay is more. It does not intend to overlook the travel industry for a reason with a shorter length. Instead, more consideration is needed to define dynamic arrangements for expanding the length of stay.
- Destination discontinuity: Expanding the average length of stay is beyond the realm of imagination in destinations with a shortage of assets and overburdened with travelers. For it, objections that are underexplored must be distinguished and grown to suit the requirements of travelers. It would help move the weight of some depleted complaints, and accessibility of assets would help expand the length of stay.

9.2.5 Development of Agro-based Tourism

Agrotourism is one of the components of rural tourism that has enormous potential for development in Himachal Pradesh. Thus, the focus of policymakers should be on the development of agro-based tourism. For it, the study provides several suggestions:

- Train and develop farmers and ranchers: To conquer the impediment of absence of aptitudes among ranchers, who are at the point of convergence of agro travel industry, preparing schools should be set up. These schools would furnish them with information about traveler conduct, building up the travel industry items, keeping up the norm of foundation, and building quality in India's travel industry.
- Provide government support for development: It was found that very little of government motivation is given till now to the advancement of agro travel industry in India. It is one of the significant reasons why the producers are hesitant to consolidate their specific specialty section in their arrangements and procedures. Without the help of the administration, this activity cannot prevail. The administration needs to give a base sum off budgetary help to improve the framework to the ranchers and guarantee that the foundation is very much kept up too. Appropriations ought to be given, and credits ought to be presented at zero financing cost. Advertising of agro places of interest ought to be crafted by the legislature.
- Generate domestic demand for the product: To handle the absence of interest of agro travel industry, the government ought to advance the travel industry at homegrown levels. Schools and universities ought to be spurred to set up visits at such destinations. Increment popularity would create the enthusiasm of the flexible side and would keep the business alive.
- Develop an innovative marketing mix: New roads in the agro-travel industry ought to be recognized. Farmers should discover better approaches to pull in the traveler by either making advancements in their item offering or developing their method of activities. It would give exceptional situating to the significant parts in the agro travel industry.
- Incorporate agro-tourism in vision and mission: Agro-tourism should have its place in the Ministry of Tourism's vision and mission. It would ensure about independent assets for the agro travel industry in the nation's spending plan. Without composed endeavors, the industry can never accomplish its objectives.
- Build up necessary infrastructure at destinations: To guarantee that vacationer inflow is normal, government must form and keep up a base degree of the travel industry foundation at such locations. Appropriate cleanliness ought to be guaranteed, well-being and safety efforts ought to be in place, hazard board systems ought to be available, and accessibility of cell phone signals, picture takers, and cafés are unquestionable requirements nowadays.
- Maintain a database of agro-tourism: An agro-traveler database could keep up to discover the positives and negatives of the industry. This information base would likewise help us in finding if the traveler was fulfilled and has visited again or not.

9.2.6 Developing Marketing Mix Suggestions

- Use of social media in marketing: As online media assumes a critical function in traveler dynamics and is essential for entire vacationer venture checking, information examination and brief activity are the need of the hour. For eatery propri-

etors, hoteliers, and transport proprietors, cost and quality are the standards they need to meet. In this way, auditors and ratings by others assume a particular function in producing drives along these lines, checking content on such channels and colluding with content networks, and advertising on pages managing such substances can do more business for them. Travel service providers and homestay proprietors need to better focus on and advertise their items. Making page on social media and better focus will help in growing/broadening the client base and enlarging the income stream; in addition, the appearance of an online travel guide and better data content open doors for partners to run business, advance crusade on the web, and trim down its expense of activities.
- Changes in products and services: Altering existing products to better suit the tourists' needs will be crucial to success for the tourism industry. It will ensure the longevity of the industry. There is a need to reassess the tourist needs and demands and modify the product and services.
- Alteration in pricing mix: Price establishes the first impression on the tourist for the destination. Therefore, pricing has to be lucrative. However, a balance has to be kept between discounts and economic benefits from the tourism industry. Thus, pricing strategies need to be made more scientifically and strategically.
- Use campaign-based promotions: As there are tourism campaigns for India and few states indulge in campaign-based advertisements, similarly, Himachal Pradesh needs to have its tourism campaign. Rural tourism must be a prominent aspect displayed in the campaign.
- Engage in a celebrity endorsement: Celebrity endorsement is also a way to engage with prospective tourists. So, the policymakers can engage a celebrity from the state with national recognition, whose appeal can attract tourists to the state. For this, a group of celebrities can also attempt to pull in different niche tourism.
- Establish customer relationship management: Customer grievance handling and feedback mechanism should be built to develop a stronger bond with the tourists. Work should be done upon their reasonable suggestions to distinguish the tourist needs and wants. Reminders must be sent on their social media profile about their last trips and memories that can emotionally persuade them to visit again.

9.2.7 Rural Tourism Suggestions

The study also highlights gaps in the rural tourism research, which could provide managers and researchers issues to ponder. It gave various approaches to invigorate the growth of the rural tourism industry. The study presents numerous available and possible channels to kick off the development in the rural tourism industry. Coupling its analysis with high-medium-low (HML) and vital-desirable-burden (VDB) analysis will help select a suitable channel that could be chosen to boost the development of the industry:

- There are plenty of revenue-generating activities in rural tourism. However, the problem is setting up the pricing of rural tourism activities (Viteri Mejía and Brandt 2015). The price should not be such high that it disinterests the tourists, and it should not be such low that it does not result in economic gains. Researchers must undertake a study to frame pricing mechanisms for rural tourism.
- Tourism is a service-based industry, where setting up standards is an uphill task. Still, researchers must identify and set up basic level standards that the destinations under the rural tourism umbrella must meet (Kumar et al. 2018b).
- The researchers could contemplate the impact of economic incentives on tourist travel motives. The researchers must also identify a set of brand ambassadors for the marketing of rural tourism (van der Veen and Song 2014). Stakeholders should provide inputs for setting up standards.
- The policymakers should develop public forums to discuss the experiences, issues, and praises of tourists visiting rural areas. They should constitute separate boards and committees for rural tourism‘s promotion and development activities (Wilson et al. 2001). It would also ensure the support of the local community, which, as a driving force, would boost the development of the industry.
- Moreover, plans made in the tourism industry are rigid. Plans oriented toward a problem become obsolete and waste when the problem is solved, thereby requiring formulating new plans (Lal and Kumar 2019). Planning in tourism should be such that it is self-sustaining in nature and can accommodate and adjust to problems and opportunities entering into and exiting from the system.
- There is cutthroat competition in the industries today. Nevertheless, there is a requirement for coordination between allied and core industries of the tourism and hospitality sector. Small businesses can play a vital role in the development of rural tourism. Consequently, there ought to be a core group or working committee comprising all entrepreneurs from all concerned sectors to better the industry (Wilson et al. 2001).
- Rural tourism destinations are underexplored and hidden throughout the country. Therefore, an exploratory study must be conducted to explore more on the destinations which could serve as tourist destinations, existing infrastructure at those destinations, and possibilities of tourism development in that region.
- The rural tourism industry cannot develop in isolation in one or two states. As rurality is available all through the nation, planning agencies could utilize its advantage to create integrated rural tourism-based circuits that connect not only rural tourism based destinations but also incorporate tourism activities such as wellness tourism, eco-tourism, agrotourism, and other similar typologies (Kumar et al. 2018a).
- By incorporating rural tourism in its vision and developing missions to achieve it, planning agencies could devote time and funds to its development. Private sector participation could be the key to success for rural tourism development (Kaurav et al. 2013). Key partnerships between the private and public sectors would bring necessary cooperation between small businesses and industrialists in the private sector and ensure efficiency in resource utilization and development (Farmaki 2016).

- With the industry getting organized, it could revolutionize the handling of monetary transactions. However, there is a lack of knowledge among the rural population, so training and technical sessions can mitigate the concerns.
- Some areas in India are underexplored, and people inhabiting these areas are unaware of the world's development. These people can benefit by interacting with the communities outside the region (Giaccio et al. 2018; Kumar and Shekhar 2020).
- A reliable, authentic, and exclusive marketing campaign is needed to promote the rural tourism industry (Fernández-Hernández et al. 2016). Moreover, through celebrity endorsement or destination branding, rural tourism marketing could be enhanced and attractive (Glover 2009). For it, the planning agencies need to increase the allocated budget for the industry.
- Since people are reluctant to change, they must be informed about rural tourism's pros and cons to gain their support (Chin et al. 2017). Researchers should also have a more profound commitment with locals' attitudes (Hadinejad et al. 2019). They must inform locals about the expected behavior and the behavior that they need to improve their hospitality. It would help them in adjusting better to the necessities and needs of travelers.

9.2.8 Improving Financial Inclusion and Social Inclusion Suggestions

The rural Himachal Pradesh falls behind in social inclusion. Further, although people have access to banking facilities and have an operative bank account, the use of primitive banking services such as online payment is limited throughout the state. There is a need to improve the financial and social inclusion. Thus, a few suggestions can be made in this regard:

- Set up microfinance institutions: There is a scope for further establishment of microfinance institutions in the state. It will reduce the reliance of rural people on the unorganized sector for borrowing and make them use these institutions' services. This will further enhance the financial inclusion in the state.
- Use correspondent banking system: In the areas where it is impossible to establish institutes, correspondent banking (CB) can increase financial institutions' density and reach.
- Enhance ease of operations using technovation: Easy payment mechanism innovations like UPI should also be encouraged in other financial institutes' products. It will enable the rural people to quickly access the products.
- Spreading education in tribal areas: Education is the key to promote social inclusion. By enhancing the literacy in rural areas, among females, literacy can be increased. The education about cultures and lifestyles of the rest of the country can be vital to increase rural areas' lifestyle. Knowledge will also broaden one's horizon to think beyond and develop.

- Improve infrastructure in rural areas: Infrastructure improvement in rural areas will also enhance social inclusion. Electrification in rural areas will lead to the advent of entertainment and communication technologies. It will also help people in rural areas to know better about what is happening in the world and adapt their lifestyle accordingly.

9.3 Study Implications

The study has implications for every stakeholder involved in the tourism industry. The following section highlights the research implications from the perspective of different stakeholders:

- Government and planning agencies: Government could achieve several of its objectives through this study. First, it is under the objective of sustainability, which the government is practicing in recent times. So, the policymakers should investigate more how rural tourism could be sustainably developed without compromising the stakeholders' needs. It can also be done by taking the views of other stakeholders while developing plans. It will make planning more inclusive, with clear-cut defined goals. Activities that can harm the environment will also be identified and listed. Stakeholders will be motivated to not indulge in such activities that harm the environment.

Second, financial inclusion and social inclusion can be achieved more effectively in rural areas through tourism. Along these lines, policymakers must identify the ways to increase the infusion within the rural regions. Setting up institutes to make tourists aware of the local customs and traditions can impart more knowledge. Financial institutions must be set up within the rural areas with limited resources. As the utilization and need increase, these institutes can be developed with improved framework. The presence of institutions will increase the curiosity to become aware of and use the facility. To increase the social inclusion effectively, social media and other technological advancements could play a boon. By managing social media pages and uploading videos of traditional dance, festivals, food items, and artifacts regularly, much information can be shared quickly and economically.

Third, generating skill-based employment opportunities has been a challenging role for the government in recent times. In this manner the investigation opens entryways for missions such as Skill India to go and target rural populations and enable them to serve as stakeholders in the tourism industry. By setting up a training center to get involved in the industry in the form of tourist guides, photographers, industry experts, or an adventurer, disguised and actual unemployment can be further reduced. As it is a skill-based work, so much income can be generated from such a segment. However, a need to regularize and certify these workers is a must, as it will act as a proof of skill and increase the trust between the parties. The generation of employment in the homeland will reduce the migration problem. It would help in preserving the local values and traditions in youth. A database for such unemployed but educated youth can be maintained to be ready for skill sharpening. By employing job enrichment and job advancement, positions can be made considerably more reassuring.

Fourth, assuming, some way or another, the private sector is attracted to work for rural tourism development, it would save a tremendous amount of government money, which could be used elsewhere. To enable the private sector's participation, easy clearance and approval facilities should be developed by the government. The private sector should also be provided with certain tax benefits or relief in corporate social responsibility if they engage in the development and maintenance of tourism destinations. The private sector's participation should also be kept under control. It might increase the overall destination cost, which would be detrimental to the industry. Quite possibly private sector participation will help to attract foreign investment in the industry. So, negotiation is a must. For it to happen, a board must be constituted with key private players, prominent industrialists, and government officials to have a dialog for the industry's development.

Fifth, the government could make an economically weaker section of the society self-reliant. With the "Atmanirbhar Bharat" campaign in full swing, the economically weaker section of the community can be given a chance to break their shackles and appreciate the fruits of the booming economy. Not exclusively would they be employed in their backyard, but also, the financial, social, and technological infusion will make their living standard rise. Sixth, the government could increment from an increase in tax revenue as most of the agriculture income from rural areas is tax exempted. Thus, for it, farmers must be taught about the benefits of paying taxes, making them educated about their tax calculation and utilizing the tax collected from them to build a smooth industry environment. When the taxpayers feel that their money is being used for their benefit, they are encouraged to pay tax fairly.

- Local population: For locals, infrastructure developed at the destination would lift their standard of living. So, the local community must enjoy the developed infrastructure without causing harm to it. They must communicate with the government the desired infrastructure and the pros and cons of developing it. The issues must be addressed on both sides, and there will be a need for a mental revolution. The local population must not compete with the tourist in enjoying the infrastructure. With increased social inclusion, locals would find it much easier to communicate with tourists and the outside world. They would be able to better know about the happenings worldwide. It would be a platform for them to show their local talent and attract millions to their social values, customs, and traditions. It would also provide them with an alternate source of earning. Large-scale disguised unemployment problem in a rural area could be tackled by developing rural tourism sustainably. So, the skilled people working in farmlands must come out to support the industry. Instead of moving to the cities searching for jobs, they should get the required training and work in their backyard. The locals must be willing to have a communicate with the authorities concerning their grievances. Whatever may be the situation, the tourists' needs and satisfaction level must not be compromised by the locals without proper assessment. A local representative must be in contact with the government. The locals must also develop trust with the tourists by advising them about the local traditions, food, destinations, customs, and religious practices.

- Tourists: For tourists, a new tourism product would be developed, unknown and unexplored. It creates excitement and an altogether new reason to visit a destination. In addition, tourists should behave responsibly and sustainably when they visit the destinations. They should attempt to utilize the resources optimally and be reasonable in their expenditure as well. They should ponder on the suggestions by the locals and make memories without costing the locals fortune.

9.4 Study Limitations and Scope for Future Research

Like every other research, current research has its limitations. The self-reported responses of the respondents are the first limitation. This approach comes with its own set of limitations. To enrich the data collection process, qualitative data-gathering techniques should be employed in addition to questionnaire-based field surveys. It will unquestionably improve the worth of the data gathered, which will enhance the findings. Further, the outburst of COVID-19 has caused severe damage to the tourism industry. The restrictions imposed and the "new normal" post-crisis will undoubtedly impact rural tourism development. The suggestions provided must be altered to suit the legal challenges posed by the disease. In addition, the psychological impact of the disease must be assessed to draw the necessary conclusions. As tourism will now be in proximity, there are chances that people might not turn to unknown destinations. All these aspects are not considered in the study. Second, the study suffers from limitations that are common to statistical analyses. There is a possibility that variation in sample size of tourists, industry experts, and stakeholders, collecting the sample, and demographic variations in the sample might lead to altogether different results. Third, the study does not acknowledge the impact of literacy and other barriers to financial and social inclusion. The exclusion may result from the other obstacles that might not be possible to overcome with the rural tourism development. Fourth, the study does not provide a means to determine the carrying capacity of destinations. The destinations mentioned as possible alternative destinations are based on assumptions and expert opinions rather than empirical findings. Fifth, with the lack of financial resources available with the government in recent times, it might not be possible to allocate optimum resources to the industry. In that case, strategic planning might not be as useful as it might seem. Sixth, there is a possibility that the model could not be generalized due to the absence of the mentioned factors at different places or for different typologies of tourism. In that case, relevant factors must be retained. New factors should be recognized and added, and the model needs to be reapplied.

From the study limitations, one can derive a plethora of opportunities for future research in allied areas. Similar studies can be conducted in states that have rural tourism opportunities. Currently, the majority of tourism is practiced in urban areas. Then, a theme-based rural tourism circuit can be developed connecting destinations.

Additionally, by enhancing the example size or applying by increasing the sample size or applying other statistical analyses, further observations could be made regarding consumer spending, their stay pattern, their consumption of services, and their demographics. Moreover, the researchers can identify different niche tourism segments to study in the state encompassing holistic development. Carrying capacity estimation of various tourism destinations can empirically suggest the alternative destinations for the exhausting destinations and products. Lastly, researchers can undertake the present study in the light of COVID-19 and determine the impact. We learned from the epidemic era that messing with nature can have negative outcomes. All parties involved must work together to ensure the sustainable development of tourism services in order for locals to welcome guests with a smile and foster an emotional bond through a compassionate demeanor. Locals' quality of life will improve as a result of less stress brought on by tourists.

References

Chin, Chee Hua, Susan Su-Zhuang Thian, and May Chiun Lo. 2017. Community's experiential knowledge on the development of rural tourism competitive advantage: A study on Kampung Semadang – Borneo Heights, Sarawak. *Tourism Review* 72 (2): 238–260. https://doi.org/10.1108/TR-12-2016-0056.

Farmaki, A. 2016. Managing rural tourist experiences: Lessons from Cyprus. In *The handbook of managing and marketing tourism experiences*, 281–298. Emerald Group Publishing Limited. https://doi.org/10.1108/978-1-78635-290-320161011.

Fernández-Hernández, Carlos, Carmelo J. Leon, Jorge E. Araña, and Flora Díaz-Pére. 2016. Market segmentation, activities and environmental behaviour in rural tourism. *Tourism Economics* 22 (5): 1033–1054.

Giaccio, Vincenzo, Agostino Giannelli, and Luigi Mastronardi. 2018. Explaining determinants of Agri-tourism income: Evidence from Italy. *Tourism Review* 73 (2): 216–229. https://doi.org/10.1108/TR-05-2017-0089.

Glover, P. 2009. Celebrity endorsement in tourism advertising: Effects on destination image. *Journal of Hospitality and Tourism Management* 16 (1): 16–23. https://doi.org/10.1375/jhtm.16.1.16.

Hadinejad, Arghavan, Brent D. Moyle, Noel Scott, Anna Kralj, and Robin Nunkoo. 2019. Residents' attitudes to tourism: A review. *Tourism Review* 74 (2): 157–172. https://doi.org/10.1108/TR-01-2018-0003.

Kaurav, R.P.S., J. Kaur, and K. Singh. 2013. Rural tourism: Impact study—An integrated way of development of tourism for India. In *Changing paradigms of rural management*, ed. R.K. Miryala, 313–320. Zenon Academic Publishing.

Kumar, S., and Shekhar. 2020. Digitalization: A strategic approach for development of tourism industry in India. *Paradigm* 24 (1): 93–108. https://doi.org/10.1177/0971890720914111.

Kumar, S., K. Attri, and Shekhar. 2018a. Strategies for holistic development of tourism industry in India: A case study of Himachal Pradesh. *Journal of IMS Group* 15 (2): 12–22.

———. 2018b. Niche tourism: Sustainability and marketing strategies. *Tourism Innovations* 8 (1): 1–9.

Lal, M., and S. Kumar. 2019. Understanding the dynamics of length of stay of tourists in India through interpretive structure modeling. *Review of Professional Management* 16 (2): 14. https://doi.org/10.20968/rpm/2018/v16/i2/141019.

Mejía, César Viteri, and Sylvia Brandt. 2015. Managing tourism in the Galapagos Islands through price incentives: A choice experiment approach. *Ecological Economics* 117: 1–11. https://doi.org/10.1016/j.ecolecon.2015.05.014.

Van der Veen, Robert, and Haiyan Song. 2014. Impact of the perceived image of celebrity endorsers on tourists' intentions to visit. *Journal of Travel Research* 53 (2): 211–224. https://doi.org/10.1177/0047287513496473.

Wilson, S., D.R. Fesenmaier, J. Fesenmaier, and J.C. Van Es. 2001. Factors for success in rural tourism development. *Journal of Travel Research* 40 (2): 132–138. https://doi.org/10.1177/004728750104000203.

Index

S. Kumar, *Sustainable Rural Tourism in Himalayan Foothills*,
https://doi.org/10.1007/978-3-031-40098-8

The manufacturer's authorised representative in the EU is Springer Nature Customer Service Centre GmbH, Europaplatz 3, 69115 Heidelberg, Germany. If you have any concerns regarding our products, please contact ProductSafety@springernature.com

Printed and bound by CPI Group (UK) Ltd, Croydon, CR0 4YY
01/10/2026
02225902-0002